# THE BRAIN

## An introduction to neurology

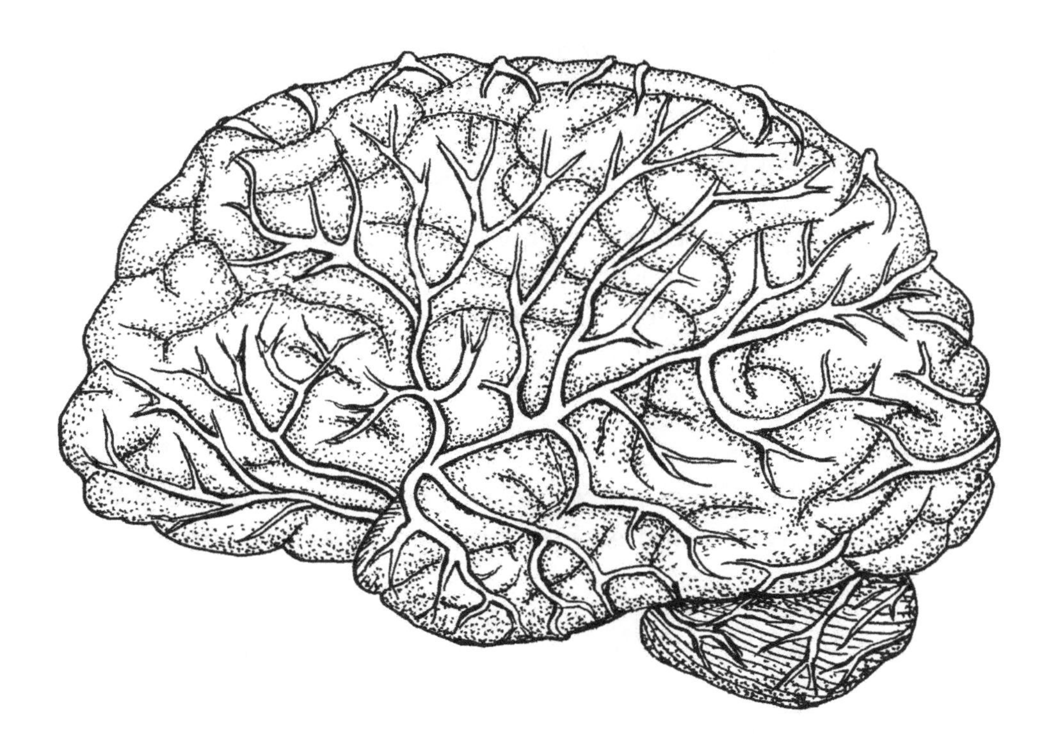

**Second Edition**

Written and illustrated by Ellen Johnston McHenry

ISBN ISBN 979-8-9868637-0-2

Published by Ellen McHenry's Basement Workshop
Pennsylvania, USA
www.ellenjmchenry.com
ejm.basementworkshop@gmail.com



**PLEASE NOTE**:
Although I have done my best to present up-to-date information on the brain, the science of neurology is growing and expanding rapidly. I encourage the reader not to rely on this booklet alone, but to use other sources of information, also.

**Nothing in this booklet should be used to diagnose any illness or syndrome.**

# THE BRAIN

## An introduction to neurology

by Ellen Johnston McHenry

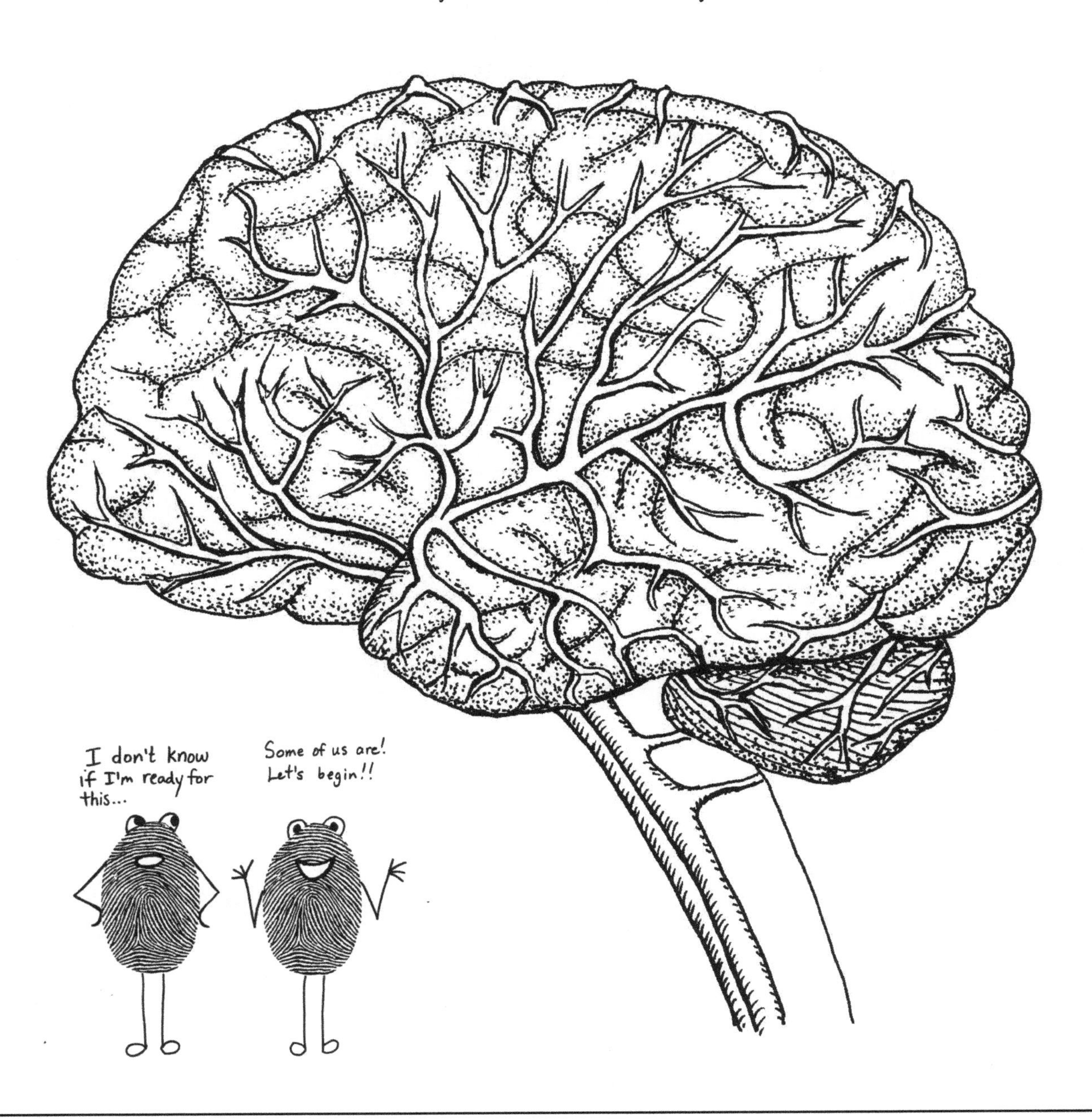

NOTE ABOUT CHAPTERS:

The chapters in this book are labeled with decimals: 1, 1.5 (one-and-a-half), 2, 2.5 (two-and-a-half), etc. The second half of each chapter has extra information that is directly related to the theme of the whole chapter, but might be just a little bit too much information for younger students. The teacher or parent can decide whether the student should be required to read the second half. For students reading the entire book, the break between chapter halves provides a natural resting point, so that the chapters don't seem so long.

# CHAPTER 1

## A VERY BRIEF HISTORY OF BRAIN RESEARCH

Even before reading this book, you already know more about the brain than the doctors and scientists of the ancient world up until about 300 B.C. If you've done even a little reading about the brain, you probably know more about it than any doctor before the 1800s. Brain research did not really begin until the late 1800s, and progressed slowly until the invention of scanning devices in the late 1900s. Even today, in the 21st century, with all our advanced medical technology, there is still a lot we don't know about the brain.

In about 300 B.C., the famous Greek thinker, Aristotle, proposed that the heart was the center of thought and emotion, and that the gray blob inside the head only served to cool the blood after it got all heated up by the heart. He was followed by a Greek doctor named Herophilus who actually cut bodies open to see how they worked. After studying the anatomy of the brain and making observations of what happened to people when they got hit on the head, he decided that the brain, not the heart, must be the control center of the body. Herophilus also noticed long, threadlike things running out from the brain to parts of the body. Today we call these things ***nerves***.

ARISTOTLE
384-322 BC

For being a "great thinker" you sure were wrong about an awful lot of stuff!

Another Greek doctor, Erasistratus, opened skulls to look at brains. He described all the wrinkles and folds that we now call ***convolutions***. He even noticed that human brains have more of these convolutions than the brains of animals do. He also saw that the brain has two main sections, and that it is covered by thin membranes. After Erasistratus, a doctor named Galen began poking around at the nerve stem that comes out of the bottom of the brain. He did experiments with animals where he cut this stem at various points and observed the results. Injuries high on the stem caused death. Injuries lower down on the stem caused only paralysis of body parts. From this he learned that the top part of what we now call the ***brain stem*** is in charge of basic functions like making the heart and lungs work.

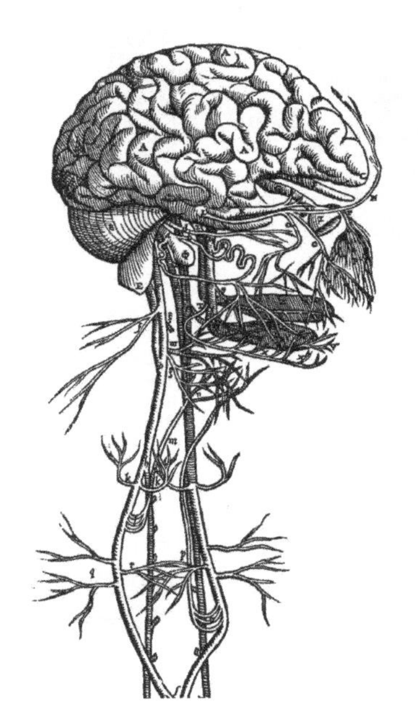

After Galen, not much happened in the world of brain science for about a thousand years. Then a doctor named Andreas Vesalius came on the scene in the 1500s. His work was the beginning of modern medical science, as he dissected bodies more scientifically than anyone had ever done before. He figured out that the brain, the spinal cord, and all the nerves in the body make up a complete ***nervous system***. His anatomy books were used by medical schools for hundreds of years. The picture shown here is one of his drawings.

Even with all of Vesalius' wonderful drawings of the brain and spine, still no one knew exactly how the brain worked. Right into the mid 1800s, doctors were clueless as to how that gray blob actually worked. Did the whole brain function as a unit, or was it subdivided into specialized areas? How did it send messages to the muscles? No one knew. That didn't stop people from guessing, though. One theory was that the brain had many specialized areas for certain activities. Someone came up with a map of the brain and labeled each area. It was thought that the areas of your brain that were large (and therefore where your talents were located) would cause your skull to bulge just a little. Thus, you could tell someone's personality by feeling the bumps on their head!

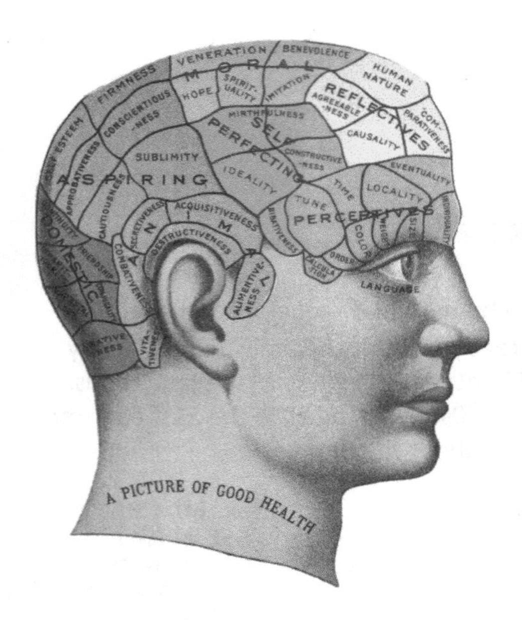


This "science" was called phrenology. Believe it or not, diagrams like this were taken seriously until the mid-1800s. If you were unfortunate enough to have a large bump on your skull right above your ear, you would be classified as a "destructive" person. If you had a bump on the back of your head, it was good news—you'd be more likely to be successful at love and friendship. A bump in front of your ear would make you eat too much. (We must wonder how many people's heads did not match their personalities!) Many of these phrenology words are rarely used today, such as "sublimity" (excellence), "suavity" ("cool factor"), "amativeness" (loving), and "approbativeness" (desire for fame).

Not all scientists believed in phrenology, however, and there was an on-going debate among scientists about whether the brain functioned as a whole or was divided up into areas. Modern science has proven that both sides in the debate were right. The brain does indeed function as a whole, and the brain does indeed have specialized locations for certain activities.

In the late 1800s, brain researchers discovered that the best way to find out how the brain works is to study brains that have been injured. After an injured patient died, they would examine the brain to find out which area, or areas, had been damaged. They discovered that in patients who had lost the ability to speak, there were always damaged cells on the left side of the brain in the area behind the temple. In patients who had lost the ability to move their arms or legs, there was always damage on the top of the brain. The researchers kept track of their findings, and began a new map of the brain, based on observation, not on wild guesses.

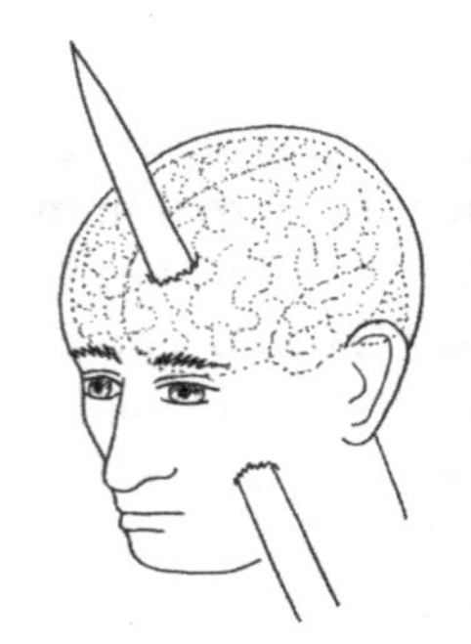

The most famous brain injury of all time happened to a man named Phineas Gage. He lived in the state of Vermont in the 1800s and worked for a railroad company as the foreman of a blasting crew. His job was to use dynamite to blast away rock that was in the path of the new railway. He would use a long metal tool called a tamping iron to carefully press gunpowder down into deep holes. One end of the rod was sharp so that it could be used to break apart stubborn clumps. On September 13, 1848, he accidentally dropped his metal rod into a gunpowder hole at the wrong time. The explosion sent the sharp end of the rod up through his head, entering under his jaw and coming right out the top. To everyone's surprise, he survived the accident and seemed to make a complete recovery. He traveled to South America and became a stagecoach driver. He became very attached to his tamping rod and took it everywhere he went. When he posed for this photograph he insisted on holding the rod.

Photograph by Jack and Beverly Wilgus of daguerreotype originally from their collection, and now in the Warren Anatomical Museum, Center for the History of Medicine, Francis A. Countway Library of Medicine, Harvard Medical School. Enlarged using Waifu2x and retouched by Joe Haythornthwaite. Own work, CC BY-SA 3.0, https://commons.wikimedia.org/w/index.php?curid=64865123

Even though Phineas had appeared to make a good recovery, those who knew Phineas said that the accident had a profound effect on his personality. They said that "Gage was no longer Gage." He was restless, had trouble making decisions, and was often rude. Before the accident, he had been such a nice young man. The rod had damaged the left side of his brain—the area containing his social skills and his ability to make decisions. Eventually, the injury did begin to affect his health and he began having seizures (sudden bursts of intense electrical activity in the brain). He died of a seizure 11 years after his accident. Years later, his skull and his tamping rod were put into a museum.

Modern brain researchers have the benefit of high-tech scanning devices that allow them to look at the brains of living patients without causing them any harm. These scans are known by their initials, **CT** (computed tomography)**, MRI,** (magnetic resonance imaging), and **PET** (positron emission tomography). The CT scan uses X-rays, the MRI uses magnetism and radio waves, and the PET scan traces the path of radioactive sugar molecules. The CT and MRI produce black and white images, while the PET scans are in color. The PET scan is used to "watch" brain activity while patients are asked to perform certain tasks. A new type of scan called the functional MRI, or fMRI, can also be used to watch the brain in action.

Modern brain surgery has also added to our knowledge of the brain. Sometimes surgeons are even able to operate on the brain while the patient is awake, which allows them to ask the patient what they feel when different areas of the brain are touched. The brain itself has no pain sensors, so this is not as bad as it sounds!

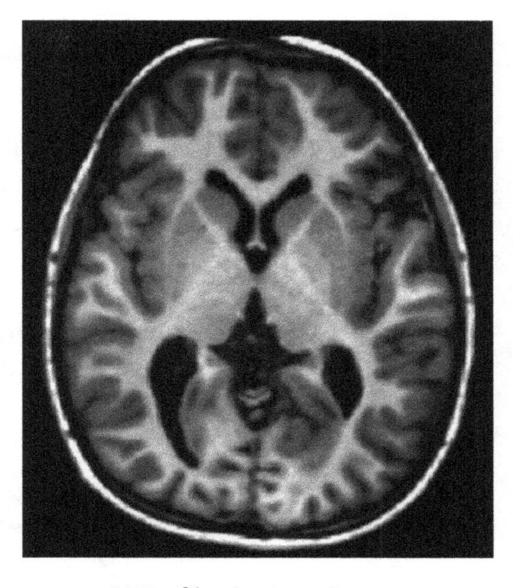

MRI of brain. top view

https://commons.wikimedia.org/w/index.php?curid=3479321

**********************************************************************************

## ACTIVITY 1.1 Use the Internet to compare types of scans

Use an Internet image search and these keys words: CT brain scan, MRI brain scan, PET brain scan. Notice the similarities and differences between these types of images.

1) Which type of scan is most similar to an x-ray and gives only black and white images? ____
2) Which type of scan seems to be used more often for side views? ______
3) Which type of scan seems the least helpful for mapping out small brain parts? _____
4) If you were a doctor, which type of scan would you NOT order if you knew your patient had metal implants in their body? _______ Why? ______________________________________________
5) If you were a brain researcher and wanted to see what area of the brain was active when mathematical calculations were being done, which kind of scan would you use? _______
Why? __________________________________________________

## ACTIVITY 1.2 Videos about brain scans

A special playlist has been set up for this book. Go to **www.YouTube.com/TheBasementWorkshop**, click on "Playlists." then find "Brain curriculum." The videos are arranged by chapter, so the first few videos are about these types of brain scans. There are also some videos about Phineas Gage.

## ACTIVITY 1.3 More about Phineas Gage

Phineas felt he was ready to go back to work only a few weeks after his accident. At first it looked like he would be able to pick up right where he left off, as the foreman of a blasting crew. His men soon discovered that he was hard to get along with and there were so many complaints about him that Phineas lost his job on the railway. While visiting his mother in New Hampshire, he was offered a job driving a stagecoach down in the country of Chile, in South America. He was promised he would not have to work with people, only horses. He got along just fine with animals, so he took the job and spent several years in Chile.

Meanwhile, Phineas' mother had moved to San Francisco. After Phineas lost his job in Chile, he went north to visit his mother. While in San Francisco, he began having seizures, which made him sick. His health started to deteriorate and eventually he had a seizure so severe that he died of it. He was buried in San Francisco, but several years later the doctor from Vermont asked to have Phineas' skull for medical research. Phineas' mother allowed the body to be dug up, and the head was removed by a surgeon. Along with the skull went the infamous tamping rod, which Phineas had taken with him everywhere he went since the accident. Both Phineas' skull and his tamping rod are now in the collection of the Harvard Medical School in Boston. (You can see these artifacts in one of the videos on the playlist.)

## ACTIVITY 1.4 Color a PET scan

PET stands for Positron Emission Tomography. To prepare for a PET scan, the patient must drink a solution that has radioactive sugar molecules in it. Areas of the brain that are active use more sugar than areas that are inactive. The PET scanner can "see" the radioactive sugar as it is used by the brain and translates this into a color image. Areas of the brain that are highly active appear red. Areas that are less active are blue or purple.

Use the numbers to color this PET image. Sharp colored pencils work best.

1 = red 2 = blue 3 = green 4 = yellow 5 = purple

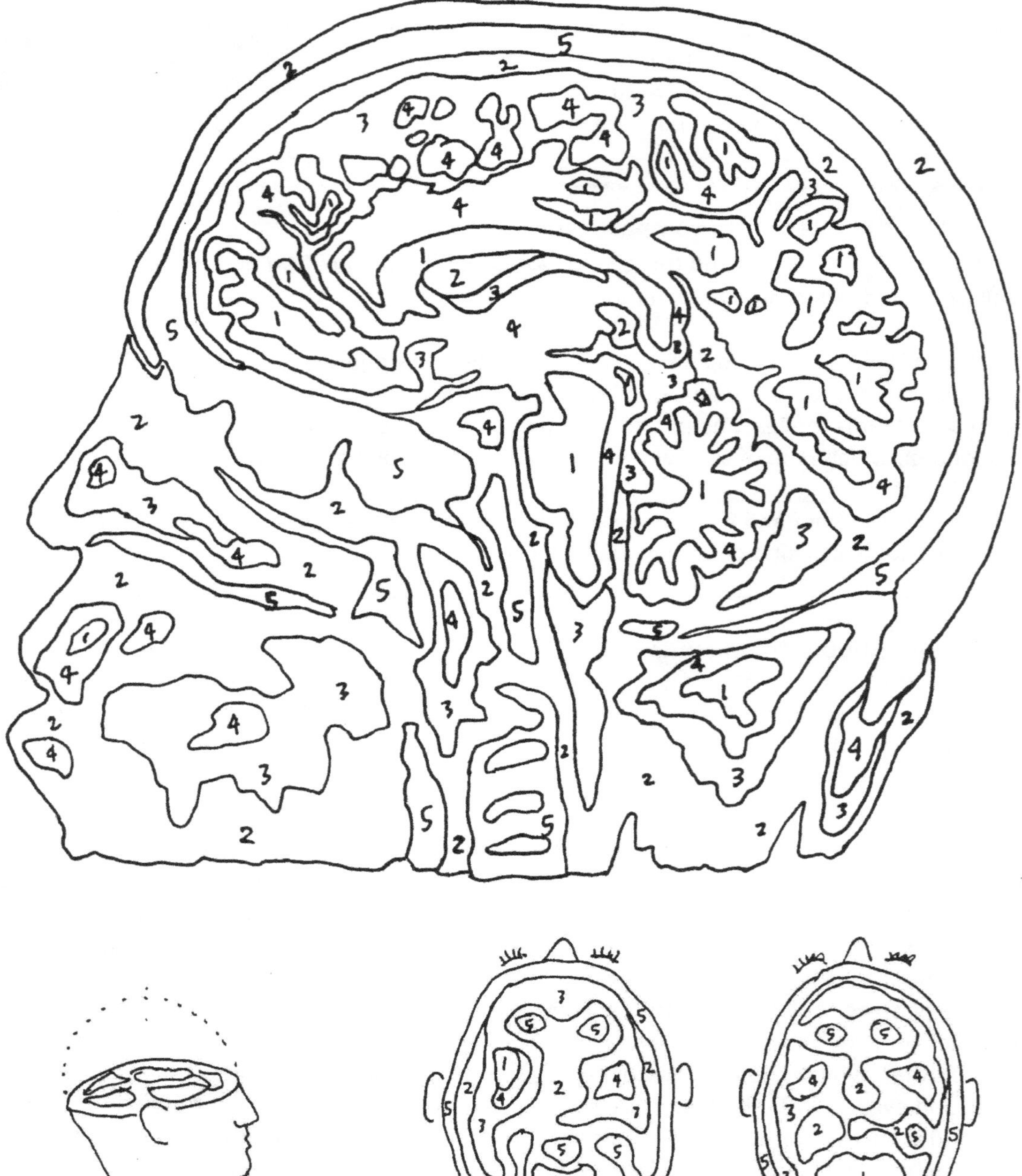

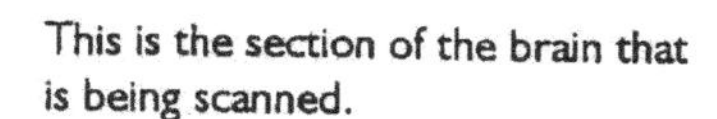
This is the section of the brain that is being scanned.

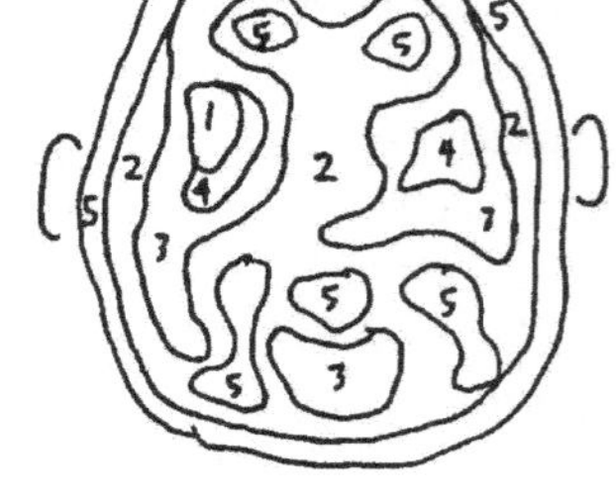
This is how the section looks when the patient's eyes are closed.

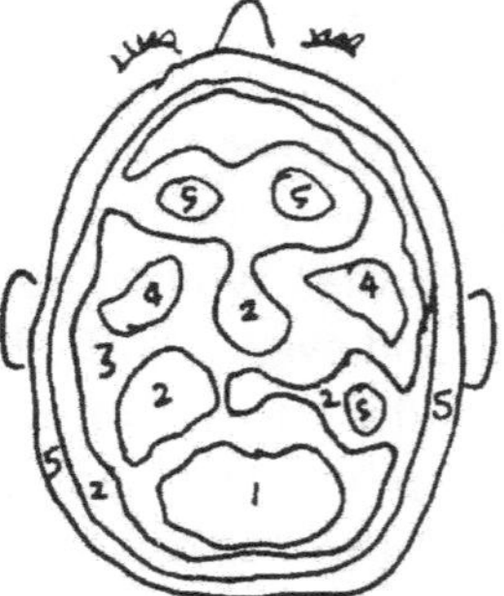
This is how the section looks when the patient's eyes are open.

# CHAPTER 1.5

## MORE ABOUT BRAIN RESEARCH

Starting in the late 1800s, doctors began to study patients with a type of brain injury called ***stroke***. In a stroke, one or more blood vessels in the brain become blocked and blood cannot be delivered to the cells that are supplied by that vessel. The affected brain cells die from lack of oxygen. Strokes are very specific, and affect only one portion of the brain; this characteristic makes them extremely useful for finding out what certain areas of the brain do. A French surgeon named Paul Broca began studying patients who could not speak after having a stroke. He asked their permission to examine their brains after they died. Enough of the patients agreed, that Broca was able to discover visible damage to cells on the left side of the brain in the area of the temple. He concluded that this area must be critical to speech. This part of the brain has become known as ***Broca's area***.

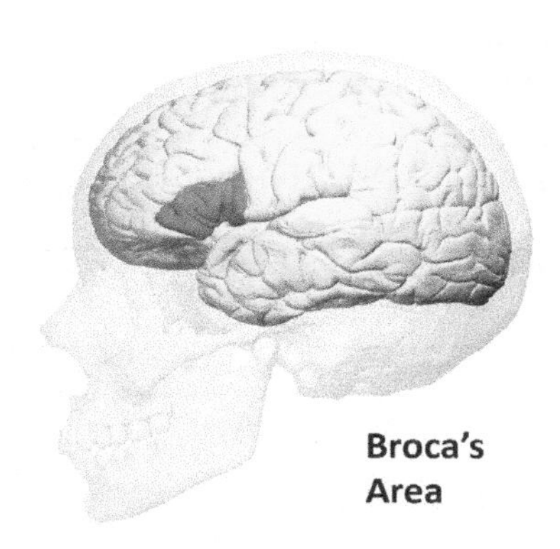

**Broca's Area**

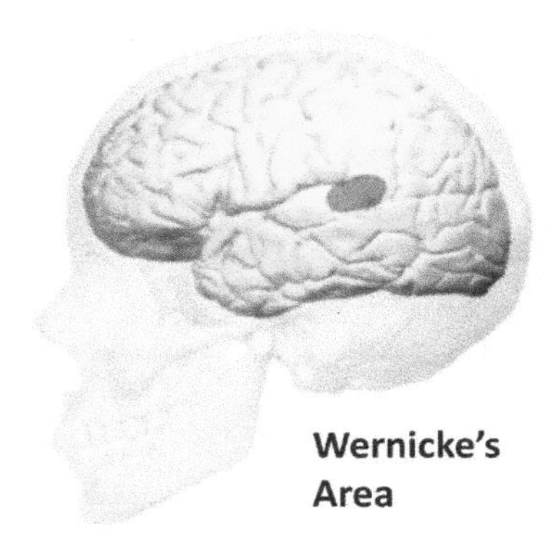

**Wernicke's Area**

In Germany, a doctor named Karl Wernicke *(VER-nih-kuh)* was also studying stroke patients. He had a number of patients who could talk, but could not understand speech. Examining these patients' brains, he found that they all had a diseased area on the left side above the ear. He guessed that this area of the brain must be vital to our ability to process the sounds and words we hear. Not surprisingly, this area of the brain has become known as ***Wernicke's area***.

As sad as brain injuries and diseases are, without them scientists would not have been able to discover how a healthy brain works. The good news is that the brain has amazing abilities to fix itself. Even in elderly patients, the brain often recovers from a stroke. Every year there is more and more evidence that the brain has more recuperative powers than was previously believed.

A fascinating story of brain research involves the brain of the famous scientist, Albert Einstein. After Einstein's died, his body was sent to a special type of surgeon whose job was to examine bodies to determine why the person died. The surgeon the autopsy secretly removed Einstein's brain and kept it. No kidding—he really did! He kept it in a jar at his own personal lab, hoping that by examining it, he could find out what made Einstein so smart. Over the years, he sent small pieces of the brain to other researchers around the world. They all came up with different theories based on their research. Some said there was no difference. One researcher found that Einstein had more glial cells (you'll learn about them in a future chapter). Others said that the area on the right side of the brain used for higher math concepts (the parietal lobe) was larger than normal. Or perhaps it was a thicker corpus callosum? We have theories, but we still don't know for sure what made Einstein's brain able to think the way it did.

Do you think he was smarter than ol' Aristotle?

Illustration credits:

Broca's Area:

By Polygon data were generated by Database Center for Life Science(DBCLS)[2]. - Polygon data are from BodyParts3D[1], CC BY-SA 2.1 jp, https://commons.wikimedia.org/w/index.php?curid=32508617

Wernicke's Area:

By Polygon data were generated by Database Center for Life Science(DBCLS)[2]. - Polygon data are from BodyParts3D[1], CC BY-SA 2.1 jp, https://commons.wikimedia.org/w/index.php?curid=32533678

## ACTIVITY 1.5 Find out more about MRI and PET

Use an Internet search engine to find more information on MRI and PET scans. YouTube or another video service could also be helpful. Answer the following questions about each.

**MRI:**

1) What does MRI stand for? ______________________________
2) How big is an MRI machine? ______________________________
3) Does it really have a big magnet in it? ________
4) Besides scanning the brain, what else is MRI good for? ______________________________
______________________________
5) Briefly describe how an MRI machine produces an image: ______________________________
______________________________
______________________________
______________________________
______________________________
______________________________
______________________________
6) What is the difference between regular MRI and functional MRI? ______________________________
______________________________

**PET:**

1) What does PET stand for? ______________________________
2) What must a person receive before getting a PET scan? ______________________________
3) What does the PET scan "see"? ______________________________
4) Why is the PET scan in color and what do the colors represent? ______________________________
______________________________
______________________________
5) Besides brain imaging, what else is PET used for? ______________________________
______________________________

## ACTIVITY 1.6 Find out more about Einstein's brain

If you find Einstein's brain a fascinating subject, there is more you can learn via the Internet. You can see actual pictures of his brain. (In fact, pieces of the brain were on public display in London in 2012.) Just use the key words "Einstein's brain" in any Internet search engine. Also, Wikipedia has article titled "Einstein's Brain."

There is a short video documentary posted on the "Brain curriculum" playlist at **www.YouTube.com/TheBasementWorkshop**. Included in this video is an interview with Thomas Harvey, the doctor who removed and kept the brain. He kept parts of the brain in various jars that spent a good deal of time in the trunk of his car or in his basement. (Rumor has it that he also removed Einstein's eyes, and gave them to Einstein's eye doctor. The eyes are now reportedly in safe deposit box in New Jersey.)

## ACTIVITY 1.7 Watch a few short videos on Broca's area and Wernicke's area

Go to the Brain curriculum playlist to see some video clips about these areas. (The exact videos available might change over time, as videos are added or removed from YouTube.)

# CHAPTER 2

## BASIC BRAIN ANATOMY

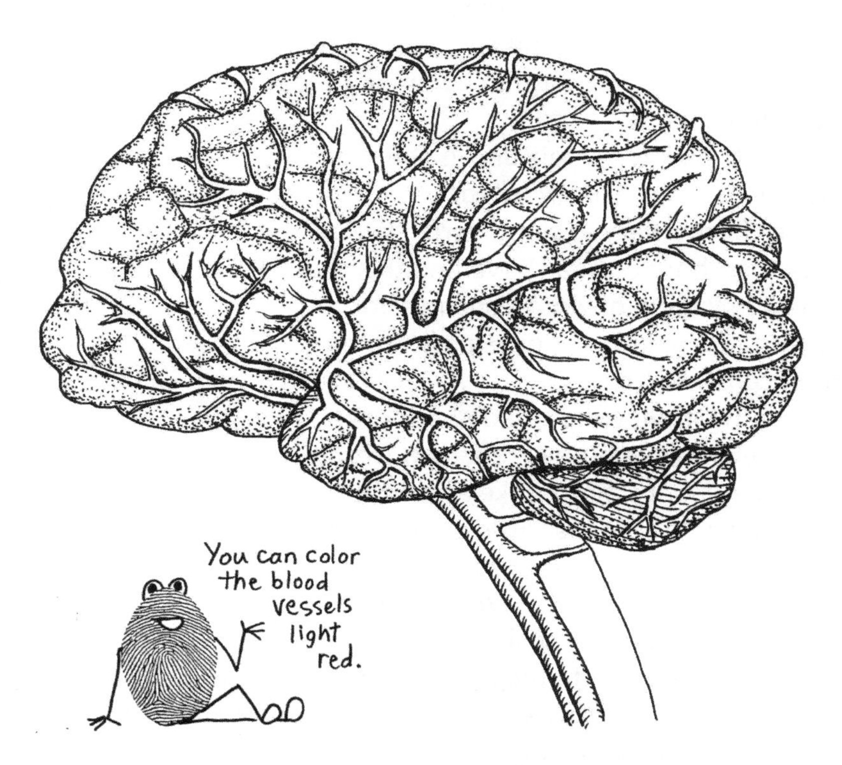

You can color
the blood
vessels
light
red.


Your brain fills up most of your head and weighs about 1.5 kg (3 pounds). The brain needs a lot of blood, so it sits on top of some very large blood vessels, almost like a ball balanced on top of a water fountain. It receives 35 liters (about 8 gallons) of blood every hour. The blood brings oxygen and sugar to the brain and carries away waste products and carbon dioxide. The brain uses more energy than any other organ of the body, consuming 40% of the oxygen and sugar you take in. ("Sugar" includes digested carbohydrate foods such as bread, cereal, rice, potato and pasta.)

The large blood vessels at the bottom of your brain branch off into smaller and smaller vessels, down to microscopic vessels called ***capillaries***. The capillaries form a fine network all through the brain, making sure that every cell gets nourishment.

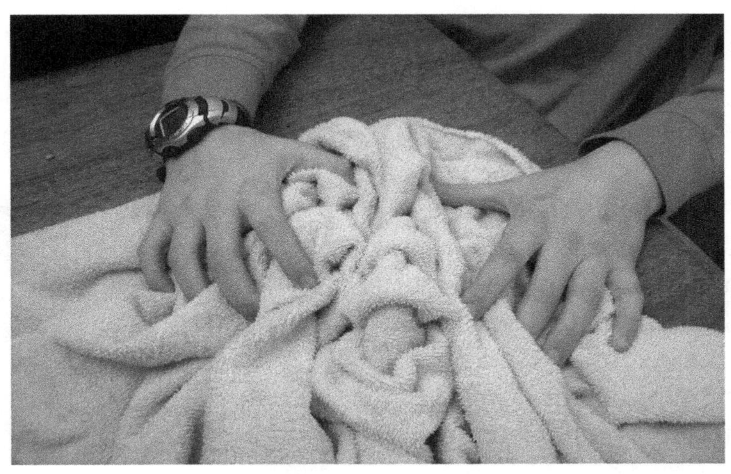

The brain looks wrinkled and folded, and there is a good reason for this. The surface layer, called the ***cortex***, is larger than it appears. If you peeled it off the brain and laid it out flat, it would cover an area about the size of a kitchen table. Imagine the cortex as a tablecloth that has been crumpled up to fit inside your head. The more wrinkles there are, the more brain cortex there is. In general, you can tell how intelligent an animal is by looking at how wrinkly its brain is. More wrinkles means more surface area, and more surface area means more space for learning and memory.

Inside the skull, the brain is surrounded by a layer of watery fluid called ***cerebrospinal fluid***, which cushions it and protects it from bumps and bangs. Although watery fluid isn't as exciting to learn about as the brain itself, it does deserve to be mentioned because without this fluid, you would be risking brain injury every time you went out to play. After the fluid bathes the brain, it flows down the middle of the spinal column, bathing all the spinal nerves. Eventually the fluid exits at the bottom of the spinal cord and is reabsorbed by other body tissues. The brain is constantly making new fluid at the rate of about a spoonful every hour.

There are three main sections of the brain. The largest section, the top part, is called the ***cerebrum***. (Most people pronounce it like *"sah-REE-brum"* but the dictionary says that *"SARE-eh-brum"* is also correct.) This is the part that you consider your "real" brain. This is where your thinking takes place. It is also the part that commands your muscles to move and processes information from your eyes and ears.

The word "cerebrum" is the Latin word for "brain." The word root "cere" comes from a very ancient word: "keres," meaning "horn or head." We find this root is a wide variety of words, such as hornet, unicorn, corn, corner, cornea, triceratops, rhinoceros, and carrot.

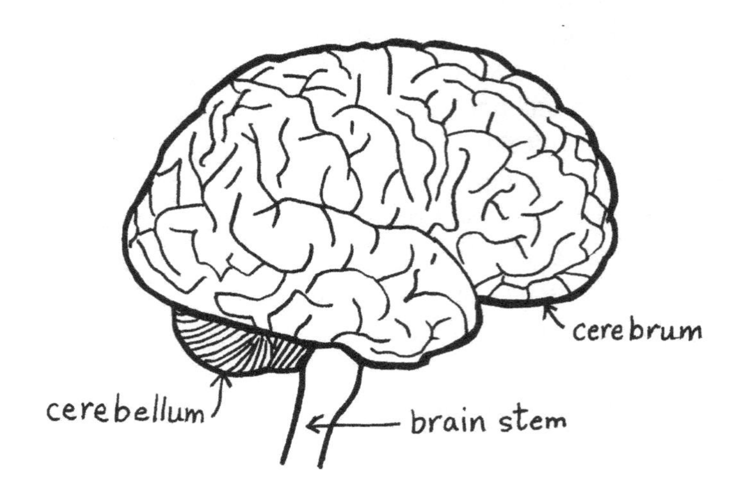


The little wrinkly blob under the cerebrum is called the ***cerebellum*** *(sare-eh-BELL-um)*. It's almost like a separate brain. Its name means "little brain." The cerebellum is in charge of coordinating balance and movement. Without your cerebellum you would fall over if you tried to walk. You need it to help you do things like throwing a ball into a hoop or shooting at a target. Also, some types of memory are stored here—your "muscle memories," such as how to ride a bike or tie your shoe. The cerebellum is the fastest growing part of the brain and reaches almost adult size by the time a child is two years old!

The stem-like thing sticking out the bottom is called the ***brain stem***. It performs the bodily functions you take for granted, such as your breathing and the beating of your heart. Some reflexes are also found here such as gagging, coughing or vomiting. Yes, these are yucky, but they keep you healthy.

Animals brains have basically the same structures as human brains. Can you identity the brain stem, the cerebellum and the cerebrum in this horse brain? Why do you think the brain stem is coming out the side instead of the bottom? (Think of what a horse's head and neck look like.) How big is the cerebellum compared to the cerebrum? Why might horses need their cerebellum to be larger than their cerebrum?

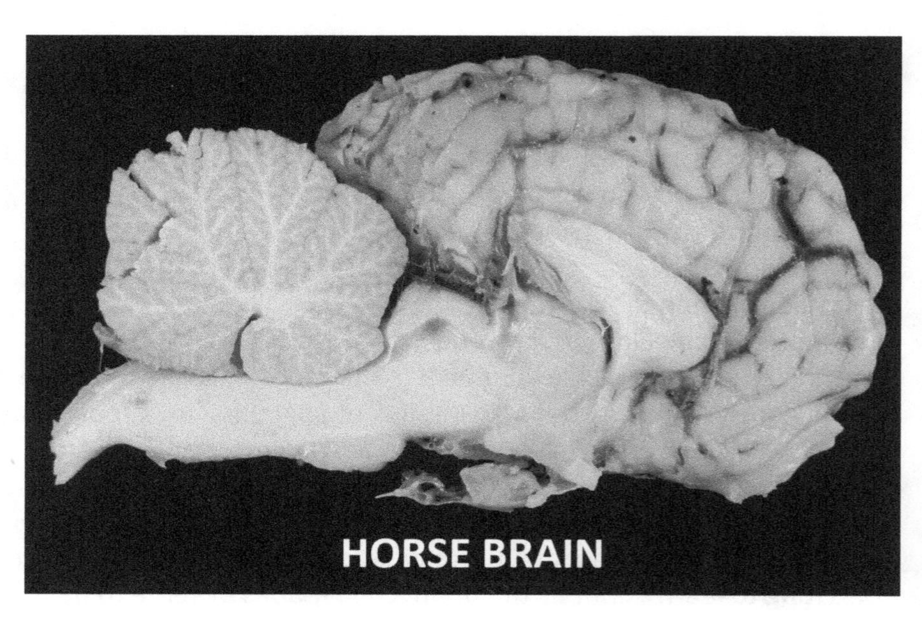

Now let's split the brain open and take a look at the inside.

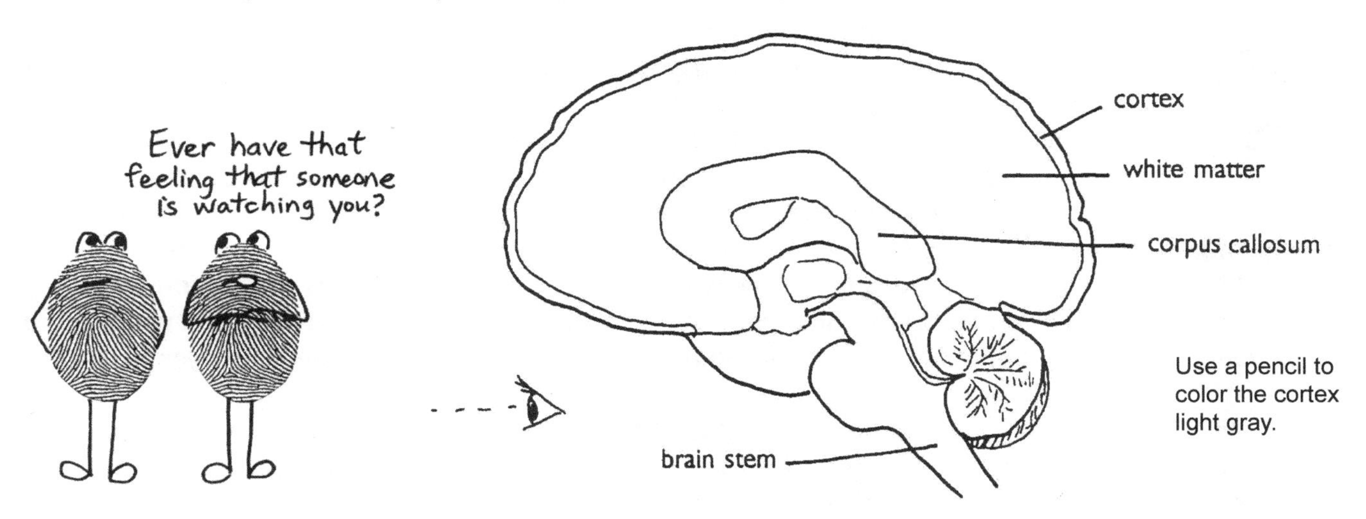


Use a pencil to color the cortex light gray.

In this diagram, we've drawn and labeled the ***cortex***. In a real brain, there isn't a line between the cortex and the rest of the cerebrum (look at the horse brain). The cortex is about as thick as a piece of corrugated cardboard. This is where thinking takes place. The cortex is sometimes called our "gray matter" because it does look light gray in color. Underneath the gray matter is a thick section called "white matter." We'll learn more about the white matter when we learn about individual brain cells and what they do. The sideways C-shaped thing in the middle, the ***corpus callosum***, is a connecting bridge between the right and left sides of the brain. ("Corpus" is Latin for "body, and "callosum" is Latin for "thick.")

If you look straight down on the brain (top view), you can see that there is a split down the middle. The cerebrum has two almost identical halves which are connected to each other on the inside by the corpus callosum. It is believed that the thicker the corpus callosum is, the better the connection between the two hemispheres. Females tend to have thicker corpus callosums, on average, than males do. You would be able to see the corpus callosum on an MRI or CT scan.

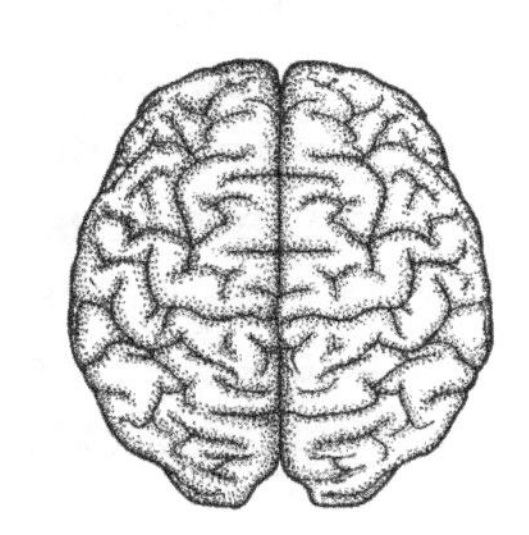

Now let's see how the cerebrum can be broken apart into six distinct areas. We know what and where these areas are because of the brain research techniques we read about in the first chapter. A real brain doesn't have any lines on it, of course. These are imaginary lines, similar to the lines between states or countries on a map. A brain surgeon must know where these areas are without any lines to help him!

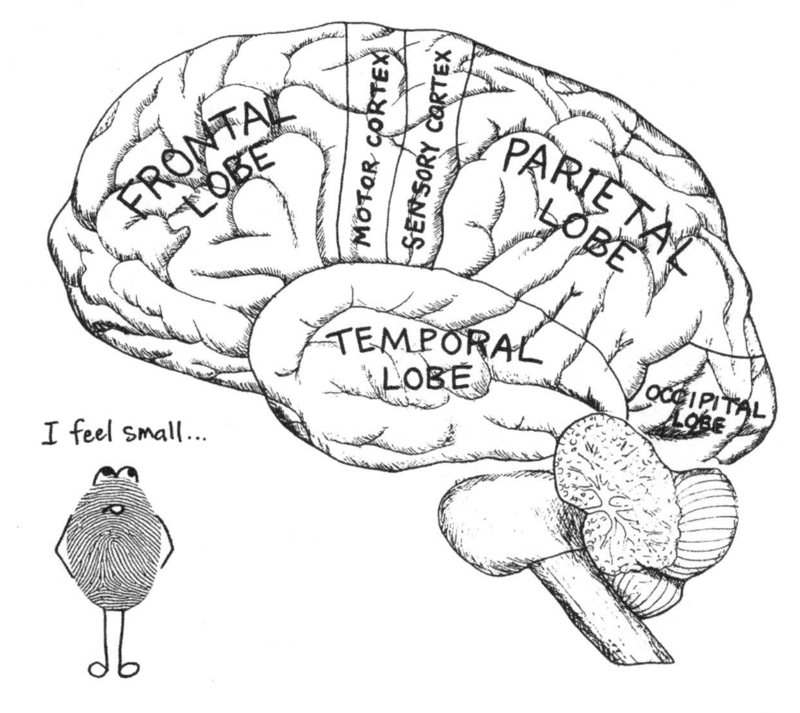


What do these six parts do? The name of the sensory cortex might give you a hint about that one. But what about the parietal or the occipital? Anatomy words are often taken from Greek or Latin so they look strange to us. However, their meanings are usually very simple.

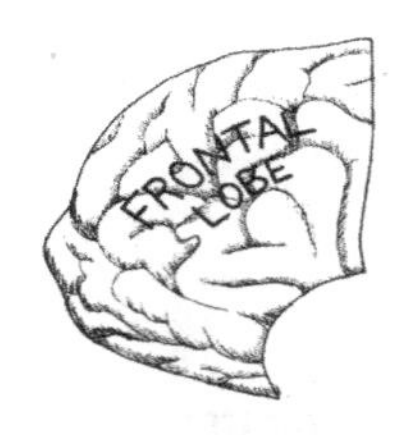


**FRONTAL LOBE:** Located in the front of your head, behind your forehead (thus its name), this is the area of your brain that you think of when you think of your "brain." (Isn't it funny that your brain can think about itself thinking?) The frontal lobe is where you make decisions and do mental calculations. When you play chess, you give your frontal lobe a real workout! The frontal lobe is responsible for monitoring your behavior and the learning social skills that you need in order to get along with other people. The famous example of Phineas Gage shows how damage to the frontal lobe can result in a change of personality. Before the accident Phineas was responsible, polite, and good with people. After the accident he was irresponsible, rude, and terrible with people. The frontal lobe has connections to other parts of the brain. For example, it gives directions to the motor cortex when you decide that you want to move your arms or legs. It also has connections to the parts that receive input from your eyes and ears.

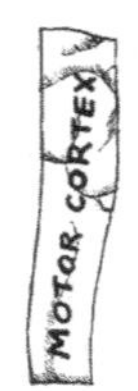


**MOTOR CORTEX:** Right next to the frontal lobe is the thin strip of cortex that sends signals to your muscles telling them to move (or to stop moving). Your frontal lobe thinks of your chess move, then sends a signal to the motor cortex, which in turn sends a signal to the muscles in your arm and hand, causing them to pick up your chess piece and move it to another square. The motor cortex also controls your legs and feet and any other muscle you can move voluntarily.

**SENSORY CORTEX:** This thin strip is connected to all parts of your body, but especially the skin. It receives messages about things you feel. Your skin has nerves that can sense hot, cold, pain, and pressure. These sensory feelings go into a middle part of your brain first, before they go to the sensory cortex. This middle part of your brain decides if the feelings are important enough to relay to your sensory cortex. This is why you don't notice minor irritations if you are very busy doing something. The sensory cortex has two sides, each one matching up to one side of the body. You might guess that the left sensory cortex matches up with the left side of your body, but the reverse is true. The left sensory cortex goes to the right side of your body, and you right sensory cortex goes to the left side of your body. Some parts of the body, such as your fingertips and face, get a much larger section of the cortex devoted to them. These parts need more sensing nerves than parts such as the back or the legs.

**PARIETAL** *(par-EYE-it-al)* **LOBE**: This area of the brain is the most mysterious one. Scientists still don't know everything the parietal lobe does. Its main job seems to be keeping track of where your body parts are. If you close your eyes, you still know what your arms are doing. You can bring your hands together without looking. This is your parietal lobe working. It also seems to keep track of all the objects in your environment, and knows "which end is up." For example, you know which end of a pencil is used for writing, even if the pencil is upside down. This may seem a bit obvious, or even silly, but this function of the brain is very important; it allows you to see objects in your mind. The parietal lobe is very important in the study of geometry and other mathematical concepts. (Remember, Einstein had a large parietal lobe.) Last, but not least, the parietal lobe works with the cerebellum and the inner ear to give you your sense of balance.

**OCCIPITAL** *(ock-SIP-it-al)* **LOBE**: This area, located at the back of your head, is where the input from your eyes gets processed. It may seem strange, but the nerves from the eyes travel all the way to the back of the brain. Not only that, but the nerves from the eyes cross over in the middle so that the left eye connects to the right side of the brain and the right eye connects to the left side of the brain. To further complicate things, the images arrive at the occipital lobe upside down! The occipital lobe must turn everything right side up and figure out what you are looking at. Your eyes only do half the job of seeing. The occipital lobe has to finish the job.

**TEMPORAL LOBE**: This area is on the side of the head, above your ears and your temples. (The temples are the sides of your forehead.) The temporal lobe is in charge of quite a few things, including your ears and nose. Your ability to speak and to understand speech is located in the temporal lobe. (Do you remember from chapter one how we first learned what the temporal lobe does?) When you talk, you are using your temporal lobe to construct a sentence that makes sense. The temporal lobe has connections to the frontal lobe and the motor cortex. The frontal lobe decides what to say and sends signals to your temporal lobe and your motor cortex. Your sense of smell and your memory of smells are also located in the temporal lobe. Smell memories are often very strong memories because the temporal lobe is located very close to the inner part of your brain where memories are stored.

*******************************************************************************************************

**ACTIVITY 2.1 A crossword puzzle about the brain words you've learned**

Clues are from chapters 1 and 2.

**ACROSS:**
3) This "firm body" is the bridge between the two hemispheres.
5) Tiny blood vessels
7) This lobe is the one you use to make decisions.
8) The ____ matter is the interior of the cerebrum. It is much thicker than the gray matter.
9) The _____ cortex receives incoming signlas from the nerves in the skin.
10) The ___ lobe keeps track of where your arms and legs are.
11) The brain ____ contains many reflexes and automatic functions like breathing.
14) The ___ lobe has many functions including smelling, hearing, speaking and listening.
15) The outer layer of the cerebrum is often called the ______.
16) This fluid brings oxygen to the brain.

**DOWN:**
1) This lobe is connected to your eyes and is responsible for interpreting visual signals.
2) This means "little brain." It coordinates movement and balance.
4) This fluid is in the "empty spaces" in the brain and helps to cushion and protect it.
6) This is the large top part of the brain. It is divided into both hemispheres and lobes.
12) The ___ cortex sends out signals to the muscles.
13) The ___ matter is the outer layer (the cortex) of the cerebrum.

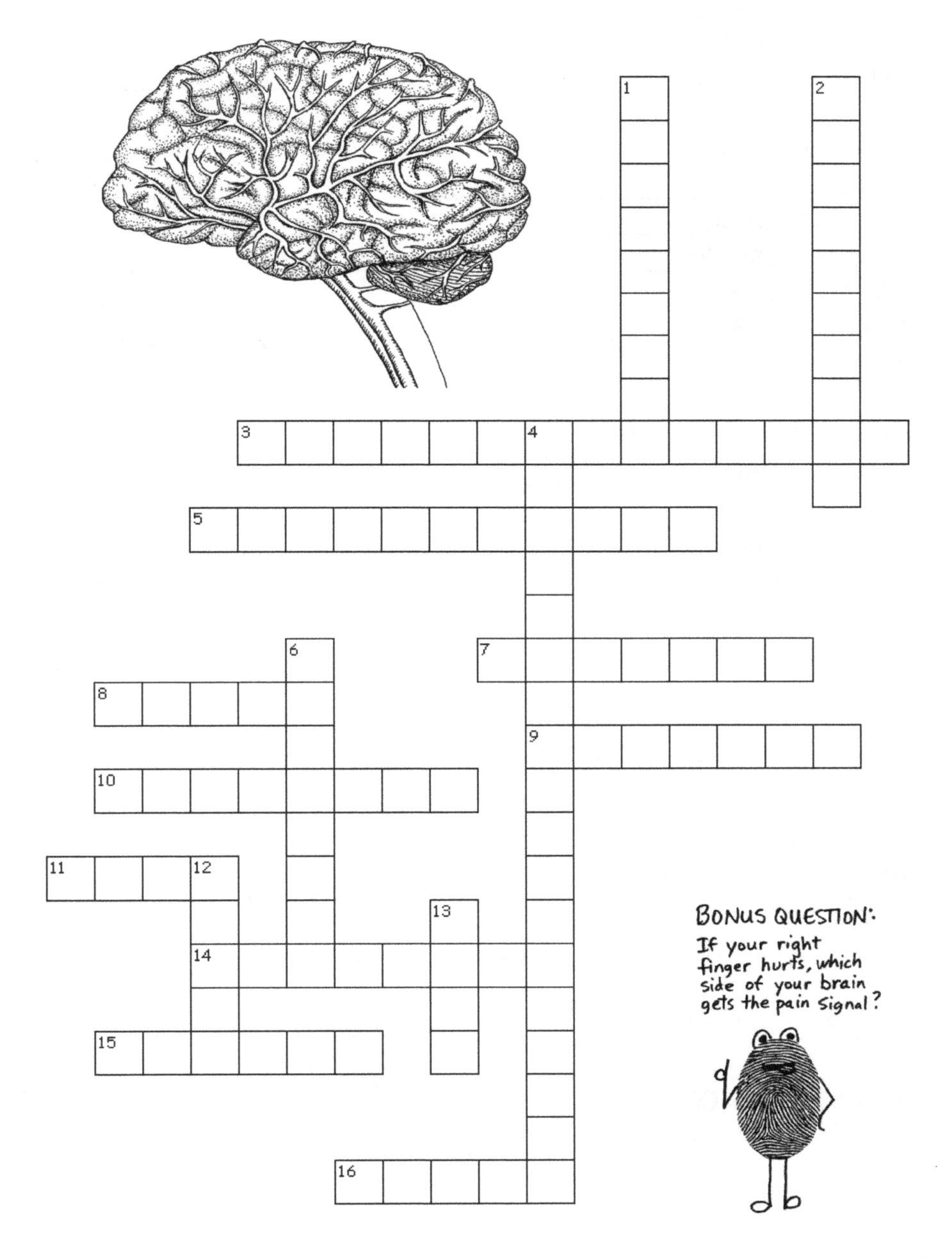

## ACTIVITY 2.2 What brain parts are used for these activities?

For each of the activities described below, list the lobes of the brain that you would use and what each one would be used for. The first one is done for you as an example.

1) Brushing your teeth:
FRONTAL LOBE: decides to brush teeth and sends signals to motor cortex
MOTOR CORTEX: tells muscles to move toothbrush around in the mouth
SENSORY CORTEX: Feels the brush in your hand, feels the scrubbing on your gums and teeth
PARIETAL LOBE: Senses that the hands are raised to the mouth, also keeps you balanced
OCCIPITAL LOBE: Makes sense of what your eyes are seeing in the mirror
TEMPORAL LOBE: Smells the toothpaste

2) Playing the piano
FRONTAL LOBE: ______________________________
MOTOR CORTEX: ______________________________
SENSORY CORTEX: ______________________________
PARIETAL LOBE: ______________________________
OCCIPITAL LOBE: ______________________________
TEMPORAL LOBE: ______________________________

3) Riding a bicycle
FRONTAL LOBE: ______________________________
MOTOR CORTEX: ______________________________
SENSORY CORTEX: ______________________________
PARIETAL LOBE: ______________________________
OCCIPITAL LOBE: ______________________________
TEMPORAL LOBE: ______________________________

4) Talking on the telephone
FRONTAL LOBE: ______________________________
MOTOR CORTEX: ______________________________
SENSORY CORTEX: ______________________________
PARIETAL LOBE: ______________________________
OCCIPITAL LOBE: ______________________________
TEMPORAL LOBE: ______________________________

## ACTIVITY 2.3 Color a "map" of the brain

Fill in a color for each square of the key, then color the corresponding brain part.

☐ FRONTAL LOBE ☐ CEREBELLUM
☐ MOTOR CORTEX ☐ TEMPORAL LOBE
☐ SENSORY CORTEX ☐ BRAIN STEM
☐ PARIETAL LOBE ☐ OCCIPITAL LOBE

# CHAPTER 2.5

## MORE ABOUT BASIC BRAIN ANATOMY

Let's take a closer look at the tissues and fluid that surround and protect the brain. These tissues are called the ***meninges*** *(men-IN-juz)* and are made of three distinct layers. The first layer, right under the skull, is called the ***dura mater***. ("Dura" means "hard or tough," and "mater" means "mother.") The texture of this layer is leathery and tough and is attached to the skull. (Mothers have to be tough sometimes!)

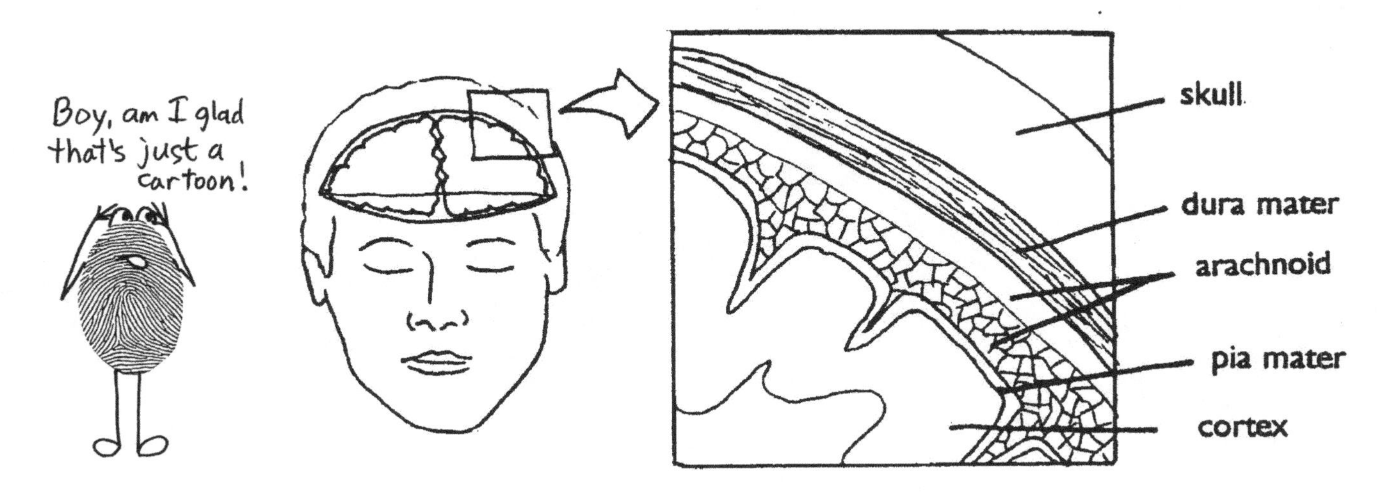


The second layer is called the ***arachnoid*** layer. This word might remind of the scientific name for spiders: the arachnids. The arachnoid layer of the brain was given this name because under the microscope it looks a bit like a spider web. It is inside the arachnoid layer that the protective ***cerebrospinal fluid*** flows.

The third layer, called the ***pia mater,*** is delicate and soft ("pia" means soft) and it attaches to the surface of the brain, following every wrinkle and bump. These three layers work together to cushion the brain and protect if from all the bangs and bumps that life inevitably brings.

Another important safety feature of the brain is called the ***blood-brain barrier (BBB)***. The tiny blood vessels in the brain (the capillaries) are close enough to the brain cells that oxygen and sugar can pass from the blood into the cells, and carbon dioxide and waste molecules can pass out of the cells and be carried away by the blood. (The brain cell in the picture is really weird-looking, isn't it? More about brain cells in chapter 5!)

The cells that make up the capillaries in the brain are joined together very tightly so that only very small molecules can get in or out. This protects the brain from potentially harmful substances. Bacteria are huge compared to nutrient molecules, so they hardly ever get into the brain. If bacteria do get in, this poses a problem for doctors because the barrier keeps most antibiotic medicines out. A sugar called mannitol is often used to temporarily loosen up the junctions between the cells and allow larger molecules, such as antibiotics, to get in.)

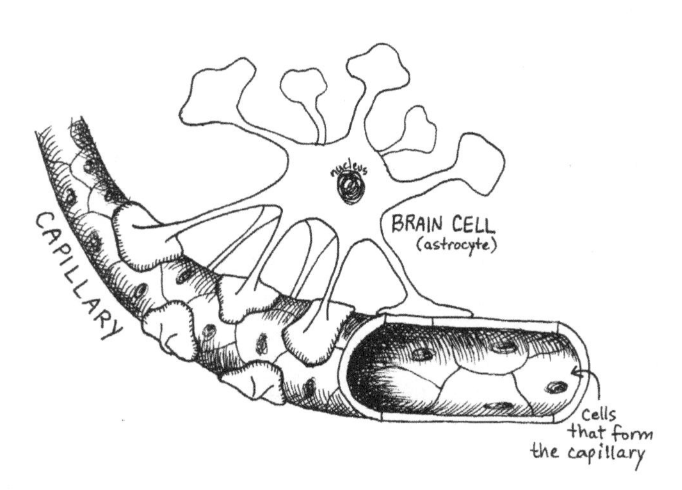


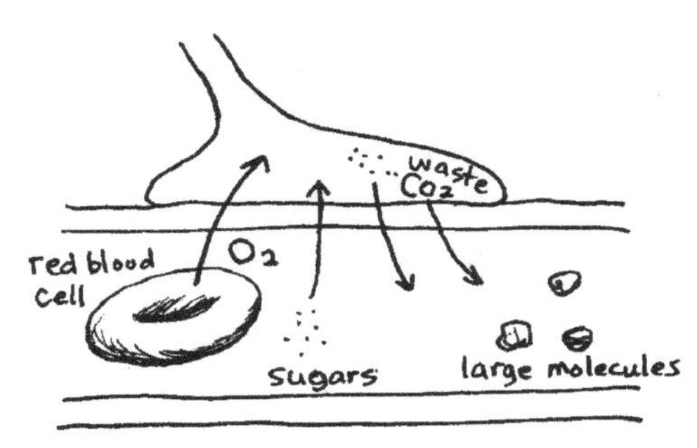

Here are a few more correct technical words for brain parts: The correct name for a wrinkly bump is a ***gyrus*** *(JIE-rus).* The valley in between is called the ***sulcus***. A split between lobes of the brain is called a ***fissure***, which is just a fancy name for "crack."

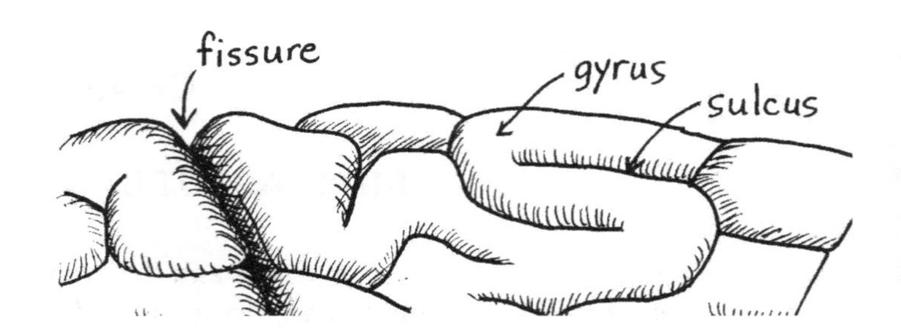


Now let's take a closer look at the sensory cortex. If you laid it out in a semi-circle and wrote down what each part of it did, you would get a "map" that looks something like this:

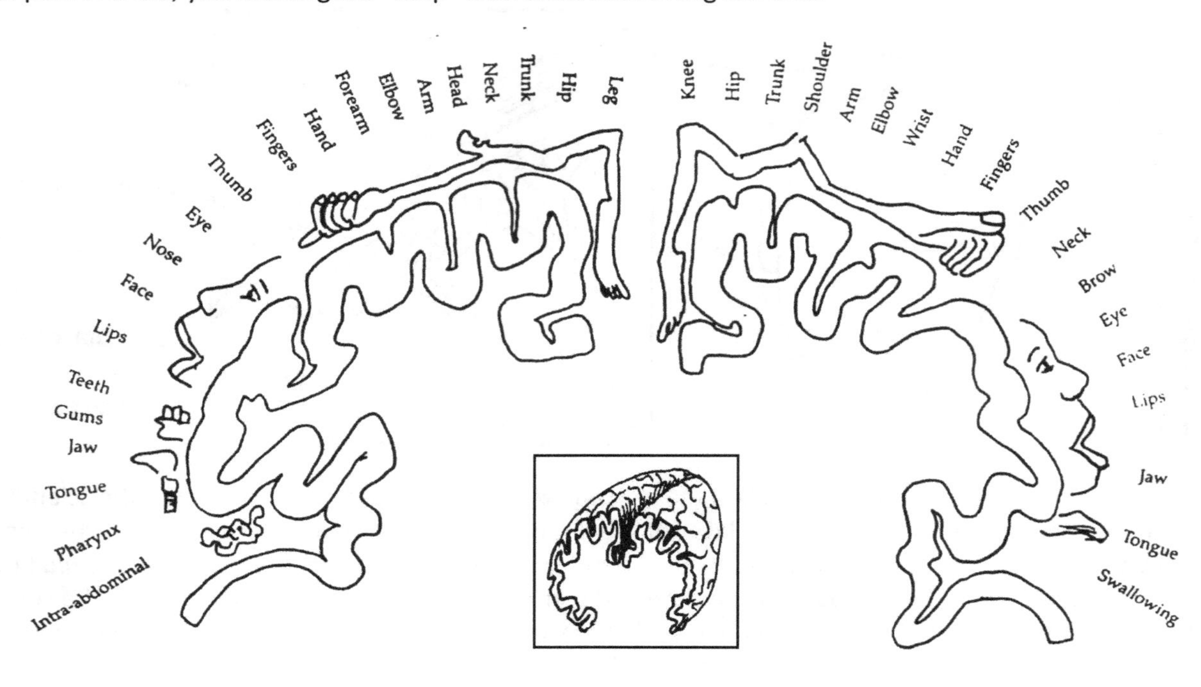


Scientists call this diagram the ***homunculus***, meaning "little man." It shows how much of the cortex (and which parts) are devoted to certain parts of the body. Your back doesn't need to be as sensitive as your finger-tips, so less of the cortex is devoted to the back. You can see that the face takes up almost as much space as the rest of the body does. The lips and mouth get a sizeable portion of the cortex. If you drew a person with these proportions, it would look very strange indeed!

******************************************************************************************

**ACTIVITY 2.4 Measure the length of the sensory cortex**

The sensory cortex is very wrinkled. You'll remember that the cortex is wrinkled so that a large surface area can be scrunched into a small space. Use a piece of string to measure the cortex in this diagram. Lay the string along the cortex, following all the twists and turns. Put a mark on the string where it reaches the end of the cortex.Then pull the string out to its full length and measure it. If you don't have a ruler with centimeters you can use inches instead. Scientists prefer measuring with metric units, as they are based on the number 10 and are therefore easier to use in calculations.

The cortex on the left is _____ centimeters long. The cortex on the right is _____ centimeters long.

## ACTIVITY 2.5 Read more about the blood-brain barrier

The discovery of the blood-brain barrier (abbreviated as BBB) goes back to the late 1800s when Paul Ehrlich, (famous for inventing many of the stains that are used in preparation of microscope slides), discovered that when blue dye was injected into a lab animal's body, it stained all the tissues in the body except for the brain. This suggested that something was not allowing the dye to enter the brain. Then it was found that if the dye was injected into the cerebrospinal fluid, the brain would be dyed, but nothing else in the body would be. This suggested that something was not allowing the dye to exit the brain. There seemed to be a barrier that would not allow chemicals to pass either into or out of the brain.

With the invention of the electron microscope in the 1960's, researchers could take a close-up look at the cellular structure of the capillaries in the brain. The capillaries are made up of individual cells, just like any other organ of the body. In the rest of the body, the cells that make up the capillary walls seem to be spaced far enough apart that almost any size or type of molecule can get in or out.

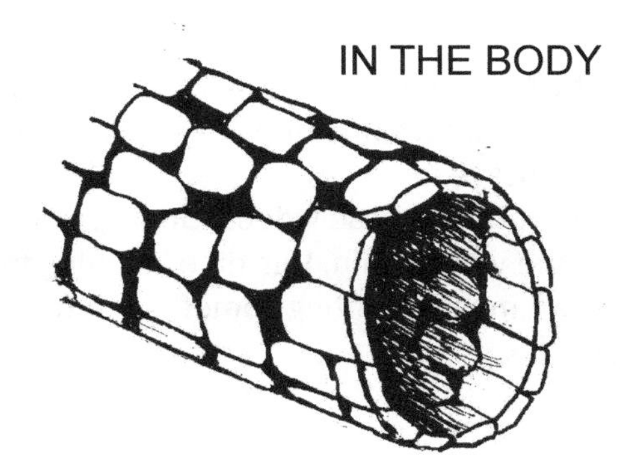


This drawing represents capillaries in regular body tissue. The cells have large gaps between them that allow molecules to go in and out easily.

NOTE: This drawing is not to scale. The cells have been simplified to make it easier to see the gaps. Real capillary cells are not square.

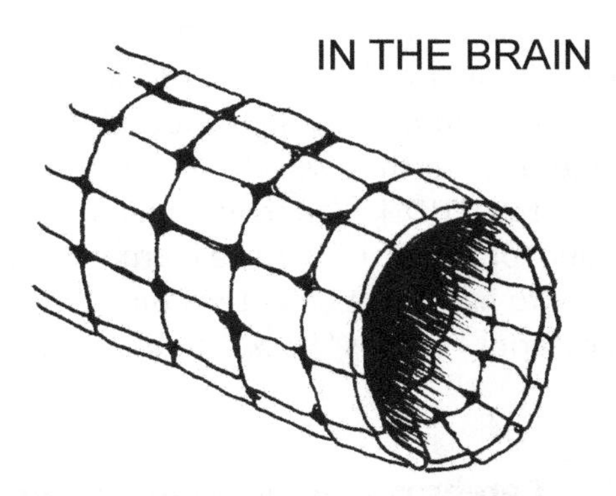


This drawing represents capillaries in the brain. The cells are squeezed tightly together so that the spaces in between are very small.

NOTE: As with the other drawing, the cells are not to scale, and have been simplified.

In brain capillaries, however, the cells are more tightly packed together. Most molecules are too big to squeeze through the gaps. The BBB blocks most molecules from crossing, except those that the brain needs (or needs to get rid of): oxygen, carbon dioxide, hormones, sugars and amino acids (the units that proteins are made of). Small molecules such as alcohol (ethanol) can also pass through the BBB. The brain does not need alcohol and yet it can get into the brain. When too much alcohol passes into the central nervous system it causes a mild case of poisoning called "intoxication." In common speech, when someone is intoxicated we call this "being drunk."

About 98 percent of all medical drugs cannot get past the BBB. This makes it very difficult for doctors to treat brain problems, especially infections. To treat brain infections, doctors must first give the patient a medicine that causes the capillary cells to loosen up. Fortunately, the BBB does an excellent job of preventing bacteria and viruses from entering the brain in the first place, so brain infections are extremely rare.

Certain diseases and conditions can weaken the BBB and increase the gaps so that larger molecules can squeeze through. The meninges are on the outside of the BBB so they are susceptible to infection. If the meninges are inflamed, the BBB is disrupted. High blood pressure can also weaken the barrier. Imagine the blood flowing through the tiny capillaries, putting so much pressure on the capillary walls that little gaps are opened up, allowing larger molecules to leak through. Diseases such as Alzheimer's and multiple sclerosis (MS) can also disrupt the BBB.

There are a few places in the brain where the blood is sampled to see how much of a certain substance is in it. The brain is the master controller of bodily secretions such as hormones, and must maintain the right balance of these chemicals in the blood. As you might guess, the blood sampling areas in the brain have a weak barrier and allow everything to pass through.

**ACTIVITY 2.6 Look at animal brains and compare them to the human brain**

Check out **brainmuseum.org**. Click on "brain sections" in the menu bar on the left. Then scroll down and click on the name of any animal to see a photographs of its brain. The charts will show you different views of the brain: left, right, bottom, top.

How do the sizes of the cerebrums compare to the total brain size? Remember that the cerebrum is where thinking takes place and wrinkles indicate a large surface area crunched into a small space. No wrinkles means less surface area, and, therefore, less thinking power. How does a rabbit brain compare to a pig? Which one would be capable of learning more tricks? Compare the manatee brain with the dolphin brain. Both are sea mammals, but which one is probably smarter? Can you spot any differences between a human brain and a chimp brain? Notice in many of the animal brains that there is a large bulbous part coming out in front of the frontal lobe. These are sensory areas, likely connected to the nose. How much of the lion's brain is devoted to the sense of smell?

**ACTIVITY 2.7 Watch some brain anatomy videos**

Use the playlist for this curriculum (**www.YouTube.com/TheBasementWorkshop**, click on "Playlists" then on "Brain Curriculum") to watch a few videos. The videos might contain some brain parts we won't cover until future chapters, but they will also review much of what you've just read.)

PANIC PREVENTION: You might hear some new vocabulary words in these videos. Don't panic! Just focus on listening for the words you've learned so far. Then, watch the video again, this time listening for new words and trying to figure out what they mean. (TIPS: "Saggital view" means splitting something into left and right sides, "posterior" means "back," and "anterior" means "front.")

**ACTIVITY 2.8 Cerebrospinal Greek and Latin**

You have noticed that most medical words are taken from either Latin or Greek. It seems like scientists are always trying to make things harder for students! Why can't they just say "membrane" and not "meninges"? Why does "little brain" have to be "cerebellum"? The reason for this is that scientists have tried to choose names that are as "neutral" as possible and don't favor any one modern European language, such as English, French, German or Spanish. Basing science words on an extinct language also guarantees that the words won't change their meaning over time. Modern Greek isn't the same as ancient Greek. The more a language is spoken, the more it changes.

This word puzzle doesn't need much explanation. Just figure out which of the Latin or Greek words on the right matches each definition on the left. For this exercise, we won't bother sorting out which are Latin and which are Greek. If you want to do that, just look the words up in a dictionary.

1) Hard: __ __ __ __
2) Soft: __ __ __
3) Spider-like: __ __ __ __ __ __ __ __ __
4) Crack: __ __ __ __ __ __ __
5) Body: __ __ __ __ __ __
6) Firm: __ __ __ __ __ __ __ __
7) Against: __ __ __ __
8) Brain: __ __ __ __ __ __ __ __
9) Little brain: __ __ __ __ __ __ __ __ __ __
10) Ring or circle: __ __ __ __ __
11) Furrow, valley or groove: __ __ __ __ __ __
12) Pertaining to the eye: __ __ __ __ __ __ __ __ __
13) Bark, covering, or shell: __ __ __ __ __ __
14) Membranes: __ __ __ __ __ __ __ __
15) Pertaining to walls: __ __ __ __ __ __ __ __

ANTI
ARACHNOID
CALLOSUM
CEREBELLUM
CEREBRUM
CORPUS
CORTEX
DURA
FISSURE
GYRUS
MENINGES
OCCIPITAL
PARIETAL
PIA
SULCUS

# CHAPTER 3

## LEFT BRAIN//RIGHT BRAIN

Have you ever heard someone say that they are "left-brained"? What are they talking about? Does that mean half their brain is missing? In the last chapter we always looked at the brain from the side. Now we are going to look straight down on the brain.

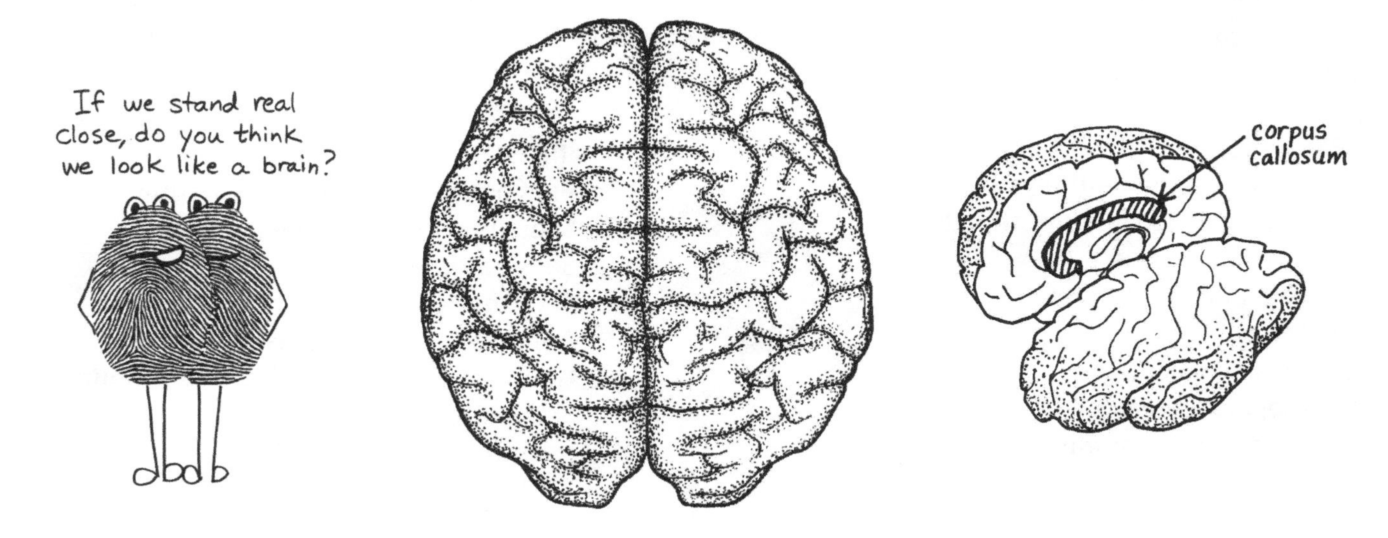


From this top view, the cerebrum looks a bit like a perfectly cracked walnut. It's wrinkly and it's got that crack, or fissure, down the center. The two halves of the brain are firmly connected in the middle by the corpus callosum (which means "firm body"). The two halves, or ***hemispheres***, look identical. They have the same lobe sections: frontal, motor, sensory, parietal, occipital and temporal. Oddly enough, however, the left hemisphere controls the right side of the body and the right hemisphere controls the left side of the body. We've already seen that the nerves from the eyes cross over and connect to the opposite sides in the back of the brain. This pattern holds true for the rest of the brain. No one is sure why this should be the case. Does it confer some kind of advantage? We don't know.

Recent brain research has turned up marked differences between the left and right hemispheres of the cerebrum, especially in the frontal and temporal lobes. We've already learned that the left temporal lobe contains important speech and language centers, Broca's area and Wernicke's area. Very rarely, someone will have their language center on the right side, instead of the left. If a brain surgeon must operate on a temporal lobe and is concerned about not harming the speech center, he can do a pre-operative experiment to find out for sure where the patient's speech center is located. The patient's right hemisphere is put to sleep for a few minutes with a mild anesthetic, then the patient is asked to answer some simple questions. If the patient can talk and answer the questions, this means that their speech center is on the left side. If they can't produce any words, then their speech must be asleep, along with the rest of the right hemisphere.

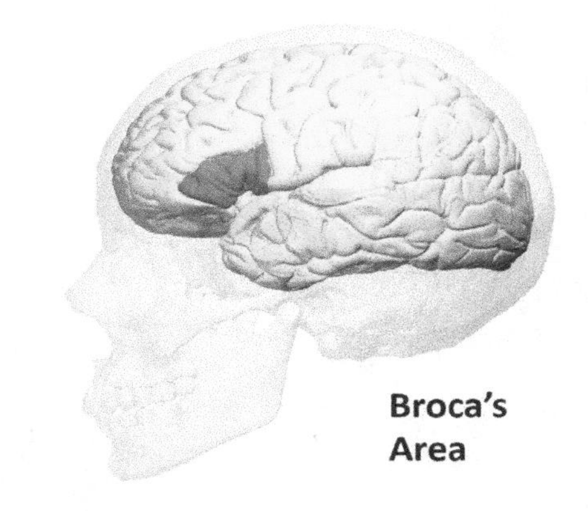

In the frontal lobe, the left side is particularly good at logic, mathematical calculations, sequencing, spelling, and vocabulary, while the right side excels in art, music, and creativity.

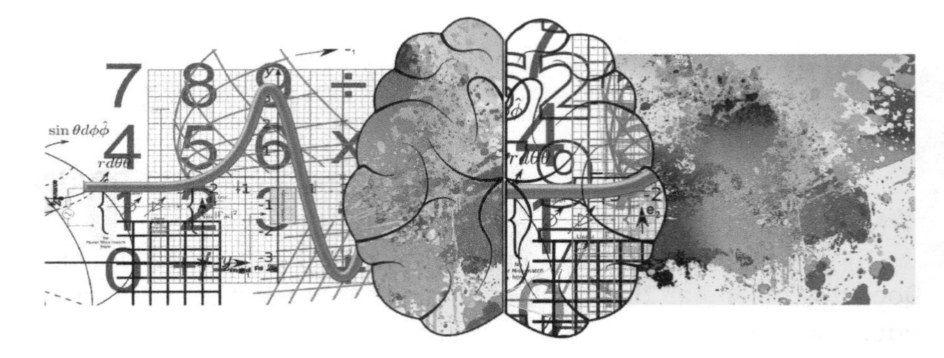

By studying people who have had brain damage on either the left or right side (and recently aided by the use of PET scans) researchers have been able to come up with a list of what the left and right sides of the cerebrum do:

**LEFT**
Uses words to name things
Uses symbols to stand for things
Keeps track of time
Understands sequences
Organizes
Counts and calculates
Uses logic and reasoning
Reads music (sheet music)

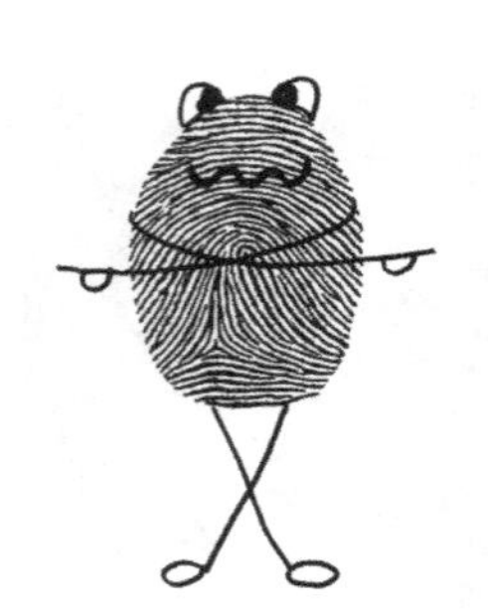

**RIGHT**
Puts pieces of information together
Sees similarities between things
Sees patterns
Sees the "big picture," not details
Appreciates and creates music
Good at drawing, painting and sculpting
Recognizes faces
Uses intuition and has "hunches"

You can see that both sides are important. The brain was designed so that the two sides would work together and complement each other. Most people use both sides of their brain about equally, though they may tend to use one side a little more than the other. When someone says that they are "left-brained," they usually mean that they are better at math and logic than they are at music and art. It might also mean that they get caught up in the details of a project and lose sight of the overall plan. People who are disorganized or have trouble keeping track of time sometimes explain it as being too "right-brained." There isn't always a correlation between hemisphere dominance and handedness. Left-handed people are not always "right-brained." Plenty of artists are right-handed.

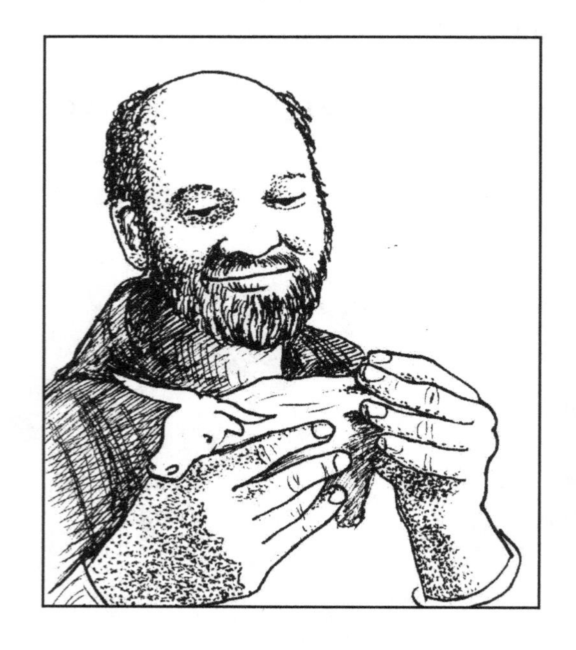

People who use one hemisphere almost exclusively are called ***savants***. They are extremely rare, and are of great interest to researchers. Savant syndrome can be caused by injury, but sometimes people are just born that way. The side they use gets extremely good at what it naturally does well. The side they don't use gets behind in its development, leaving the person with disabilities. There is a famous case of a man from Colorado, USA, whose left hemisphere was damaged by an accident in childhood. As he grew up, the right side overcompensated for the loss of the left side and he developed into a sculpting genius. He is so talented that he can sculpt any animal perfectly after seeing it only once. If you look at the "left side" list above, you will be able to guess what he can't do. He can barely talk, and he can't read, write or count. He must rely on other people to help him with basic life skills like cooking, cleaning and making appointments.

### ACTIVITY 3.1 Meet Alonzo Clemens and other savants

If you go to **www.YouTube.com/The BasementWorkshop**, you can watch a short video about Alonzo Clemons, the sculpting savant from Colorado. There are also videos about other people with savant syndrome. Remember, "savants" are not just people who are good at what they do. They have genius qualities in one area, but they also have serious deficits. Kim Peek, for example, has almost perfect recall of everything he reads (and he reads a lot!) but is mentally disabled in other ways and will always need someone to take care of him. The most fortunate savanat is Daniel Tammet, who suffered seizures during his early childhood. The seizures left him with autism, but also with amazing abilities to recognize patterns in math and in language. As a young adult, Daniel was able to learn social skills and oversome much of his autism. He can now talk about what it is like inside his head, how he literally sees numbers and simply "reads landscapes" in order to do a calculation.

### ACTIVITY 3.2 Fighting for dominance: witness a struggle between the hemispheres!

The two hemispheres usually cooperate by taking turns being in charge. These turns can last a split second, a few minutes, or several hours, depending on what we are doing. Both hemispheres are still active, but one is the boss. In this activity you will actually watch the hemispheres taking turns every few seconds.

Put this page about 2-3 feet (50-80 centimeters) in front of you. Relax your eyes and cross them until you see a third (somewhat blurry) circle in the middle with the blue and red circles right on top of each other. If you stare at this overlapped image, you will see it alternate back and forth from red to blue to red to blue. The white line in the middle will go back and forth from horizontal to vertical, along with the color change. This switching back and forth is your left and right hemispheres taking turns being dominant. When blue is on top, your right side is dominant. When red is on top, the left side is dominant. Remember, the eyes are attached to the opposite sides of the brain.

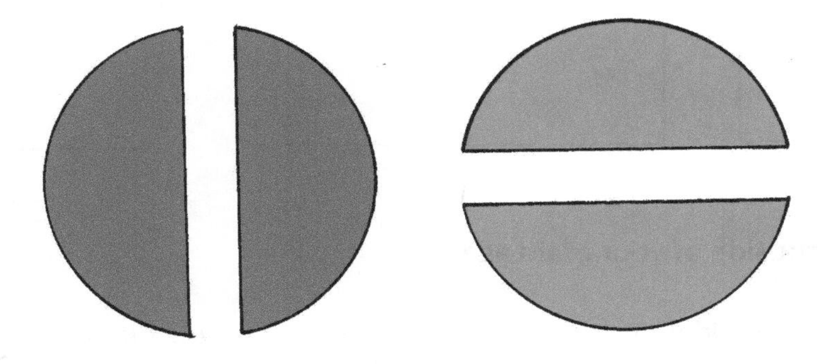

### ACTIVITY 3.3 Use the left side of your brain

Can your left hemisphere figure out what these place names are? (places in the USA)

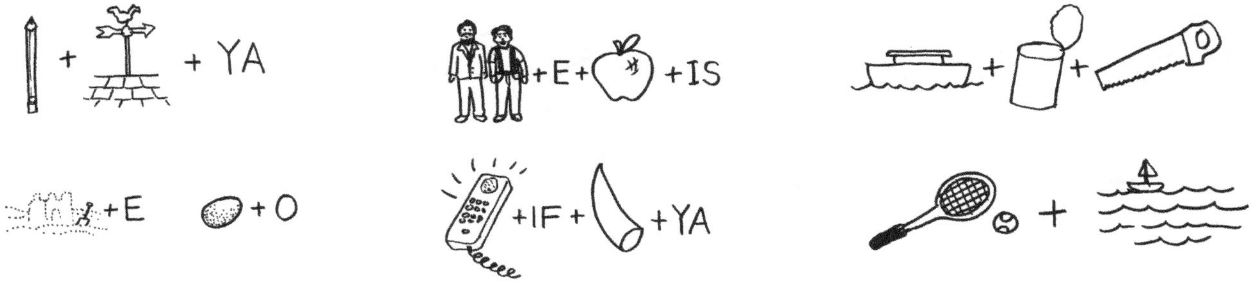

+ + YA    +E+ +IS    + +

+E  +O    +IF+ +YA    +

**ACTIVITY 3.4 Use the left side of your brain again**

Make an interesting-looking squiggle in each of these boxes:

**ACTIVITY 3.5 Use the right side of your brain**

Now make an exact copy of each squiggle in each of these boxes. You will be drawing using the right side of your brain. You'll have to look carefully at the originals, above.

**ACTIVITY 3.6 Use the right side of your brain again**

Three of these figures are identical, even though they are pointing different ways. One of them is reversed left to right, making it different. Which one is different? You will have to use the right side of your brain to rotate them in your mind.

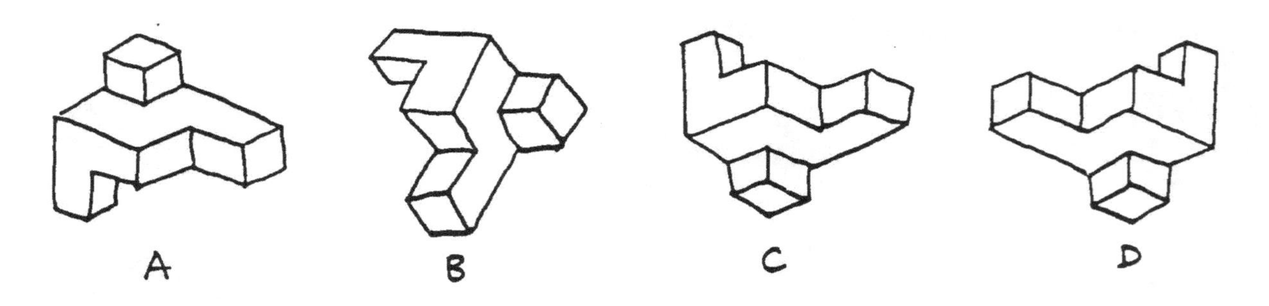

# CHAPTER 3.5

## MORE ABOUT LEFT BRAIN/RIGHT BRAIN

What would happen if you cut the corpus callosum so that the hemispheres could no longer cooperate? Any volunteers to try it? Of course not. No one would volunteer for such a drastic experiment! Fortunately for brain researchers, there are a few people out there who absolutely had to have their corpus callosum cut in order to save them from a deadly problem with their brain (such as severe epilepsy). After they recovered from the surgery, they allowed researchers to study their rare condition. These patients seemed to be normal. They could walk, talk, read and play sports much the same as they had before their operation. However, if researchers showed a picture of an object, such as a spoon, to the left eye only, the person could not say what it was because the nerves from the left eye travel to the right side of the brain, and the right side does not have language ability. If they showed a picture of a spoon to the right eye only, the person could say, "Spoon," but they could not pick out the spoon when given a selection of objects on a table. This experiment showed that the left side can name objects, but cannot match shapes. The right side can match shapes, but cannot say names. It is the corpus callosum that tells each side what the other side is doing.

This statue was inspired by a painting by Renaissance artist Arcimboldo. The left side of your brain sees fruits and vegetables. The right side sees a face.

In a similar experiment, researchers asked these "split-brain" patients to look at a picture of a face similar to the one shown here. Obviously, one side looks like a woman and the other looks like a man. The patients were asked to focus on that dot you see there on the forehead. While the patients were focusing on that dot, the researchers asked them to say out loud what they were looking at. Since the left hemisphere contains the speech center, the left hemisphere would need to be dominant in order to speak. Therefore, it would focus on the image coming from the right eye (the nerves cross over, remember) and the patient would say out loud, "I see a man." But if they were then shown another page, with two faces on it, a normal male face and a normal female face, and were asked to point to what they saw, they always pointed to the female face. (You can see "split-brain" patients doing these experiments if you go to the YouTube playlist for this curriculum, or use key words "split brain experiments" in another online video service.)

Very few of us have had our corpus callosums cut. Even if we tend to use one side a little more than the other, we still use both sides. However, some of us have one hemisphere that seems to be dominant most of the time. This can make learning a challenge if we are put into an environment that favors the side we don't use as well. Researchers have determined that without intending to do so, most traditional school programs tend to favor the left side. This puts right-side-dominant learners at a disadvantage. For example, right-side-dominant students will often struggle with spelling and seem unable to learn how to spell using traditional methods. Very few educators will figure out why these students aren't doing so well, or will be willing to make changes in the classroom to adapt to right-brain learners. Fortunately, the situation is better now than it used to be, as more teachers are learning about hemisphere dominance, and are trying to find ways to make their classrooms more "right-hemisphere-friendly."

**ACTIVITY 3.7 Do you have a dominant hemisphere?**

Put a check mark on the side that is more true. If they are both equally true, or if neither of them is true, put a check in the middle. Try to put as few checks in the middle as possible. (NOTE: This activity could be difficult for younger students. You may not have enough life experience to have figured yourself out.)

| | | |
|---|---|---|
| ___ I would rather do a report than a project. | ____ | I would rather do a project than a report. ___ |
| ___ When I listen to someone talk, I don't see any images in my mind, I just hear words. | ____ | When I listen to someone talk, I often think of visual images of what they are saying.___ |
| ___ When I read, I love to remember lots of the details because they are important to me. | ____ | When I read , I just like to get the main idea and not worry too much about the details. ___ |
| ___ When I talk, I don't usually use hand gestures, I just talk. | ____ | When I talk, I often use my hands and arms to illustrate what I am saying. ___ |
| ___ On tests, I prefer multiple choice or true/false questions, not essay questions. | ____ | On tests, I prefer essay questions, not multiple choice or true/false.___ |
| ___ It really bugs me if things are out of place in my room. | ____ | My room is seldom really clean, but it just doesn't bother me. ___ |
| ___ I do my best thinking while sitting up. | ____ | I do my best thinking while lying down. ___ |
| ___ I like the arrangement of furniture in my room to stay the same, if possible. | ____ | I like to rearrange my room often. ___ |
| ___When learning a dance step, it is easier for me to learn the sequence of moves first, talking my way through them. | ____ | When I learn a new dance step, I just like to imitate the teacher. ___ |
| ___ I can tell how much time has passed, even without looking at a clock. | ____ | I can easily lose track of time. ___ |
| ___ When I get answers to math problems, I can explain how I got the answer. | ____ | I can get the answer to a math problem, but I cannot explain how I got it. ___ |
| ___ I like to have fun without taking risks. | ____ | It's fun to take risks. ___ |
| ___ I don't mind following someone else's plans if they are explained well. | ____ | I don't like following plans. I like to make my own plans. ___ |
| ___ I can figure out what will happen next. | ____ | I can "sense" what will happen next. ___ |
| ___I spread my work out evenly over the given time I have to do it. | ____ | I don't mind waiting until the last minute to get my work done. ___ |
| ___ I like to read the instruction manual before I try to put something together. | ____ | Instruction manuals just confuse me. I do better without them. ___ |

To score this test, you simply add up the number of checks on the left and the number of checks on the right. Whichever is greater is the hemisphere that is more likely to be dominant.

# CHAPTER 4

## DEEP INSIDE THE BRAIN

We've talked about the basic anatomy of the brain and haven't even mentioned functions such as memory and emotions. Obviously, there must be more parts to the brain. In this chapter, we will take a look at the parts that lie deep within the center of the brain.

The center of the brain is a confusing place. There are all sorts of blobs in weird shapes. It's even more complicated than this picture. We've tried to simplify it as much as possible without leaving out any major parts. There are also connecting tissues, empty spaces, and a few other small parts that have difficult Latin names and serve mainly as helpers for these main parts.

NOTE: This drawing shows the left side of the inner brain. To see some 3D picture of brain anaomty, go to www.neuroanatomy.ca and click on "3D' then on "3D neuroanatomy models." The pictures spin around so you can see all sides.

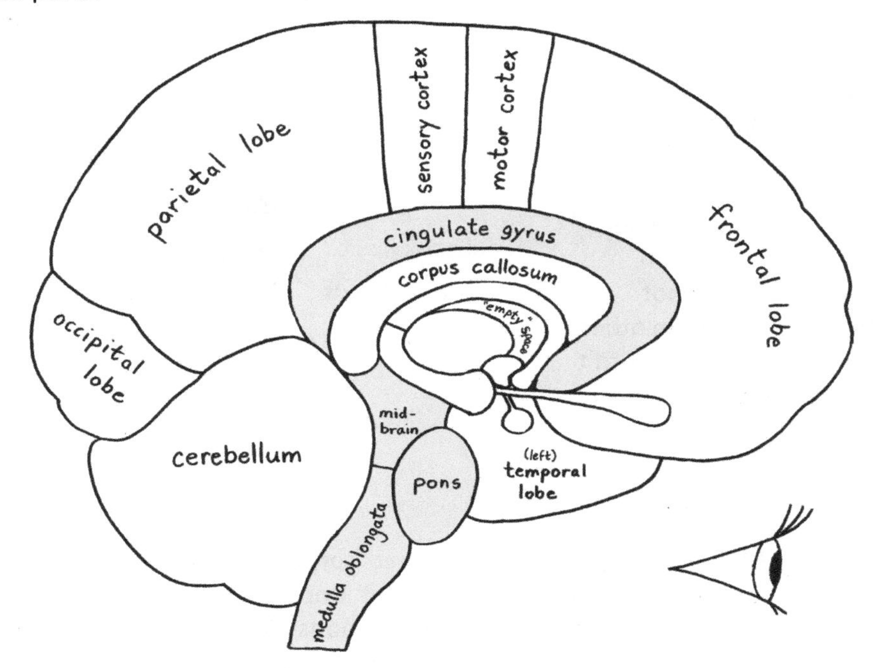


Once again we see the cerebellum and the parts of the cerebrum: frontal, motor, sensory, parietal, occipital and temporal. The left temporal lobe is sort of peeking out from behind the cerebrum. The right temporal lobe, the one that would have been in front, had to be removed in order to see the middle of the brain.

We've already learned about the corpus callosum. It's the "firm body" that joins the two halves of the cerebrum, and lets each half know what the other half is doing. Lurking above (and sort of on the outside of) the corpus callosum is the ***cingulate gyrus*** (SING-gew-late JIE-rus). It is actually part of the cerebrum, but it is attached to the ***limbic system*** (the inner parts that we'll learn about on the next page). It may help relay information from the inner parts of the brain out to the cerebrum, as well as helping to control your emotional responses.

The brain stem has been divided into three pieces: the ***medulla oblongata***, the ***pons***, and the ***mid-brain***. The ***medulla oblongata*** controls automatic funcctions like breathing, blood pressure, and the beating of the heart. Your medulla keeps these systems running even when you are asleep. The actions of sneezing, coughing (the kind you can't control) and vomiting are also in the medulla. Some cough medicines (the suppressants) act on the medulla, dulling it a bit, giving you some relief from the urge to cough.

The ***pons*** controls your waking and sleeping cycles. Even if you don't set an alarm clock, you will still eventually wake up, thanks to your pons. It also acts as a connector between the cerebellum and the cerebrum and so may help with balance and posture.

The ***mid-brain*** is sort of a "crossroads" for nerve pathways. It helps direct in-coming signals and routes them to the correct area of the cerebrum. It seems to help in the processing of information coming in from the eyes, ears and muscles. It also seems to have a function in swallowing and salivation reflexes, as well as helping to regulate sleep cycles and body temperature.

Now we will look at just the very center of the brain. This area is sometimes called the "lower brain" because it is below the cerebrum. Some of these parts work together to form the ***limbic system*** (hippocampus, fornix, amygdala) which controls emotions and is also connected to memory. If you search the Internet for diagrams of the limbic system or the lower brain, you will find that no diagram exactly matches any other. Few of them will agree on the shapes and sizes of the parts, and they will even use different labels than the ones we will show here. Brain parts are not as distinct and easy to identify as bones or digestive organs. The brain is a mass of soft, squishy, lumpy gray stuff, and, for the most part, the divisions between parts are not easy to see. Scientists can disagree on where the divisions are or whether some parts are just sub-sections of other parts. And just to confuse you further, they can't agree on names. (For example, the pituitary gland can also be called the hypophysis.) This makes learning lower brain anatomy very difficult. We've tried to make it as simple as possible, but please be aware that the diagram you see here might not match the diagrams in other textbooks. (In fact, it doesn't even match our own drawing on the previous page!)

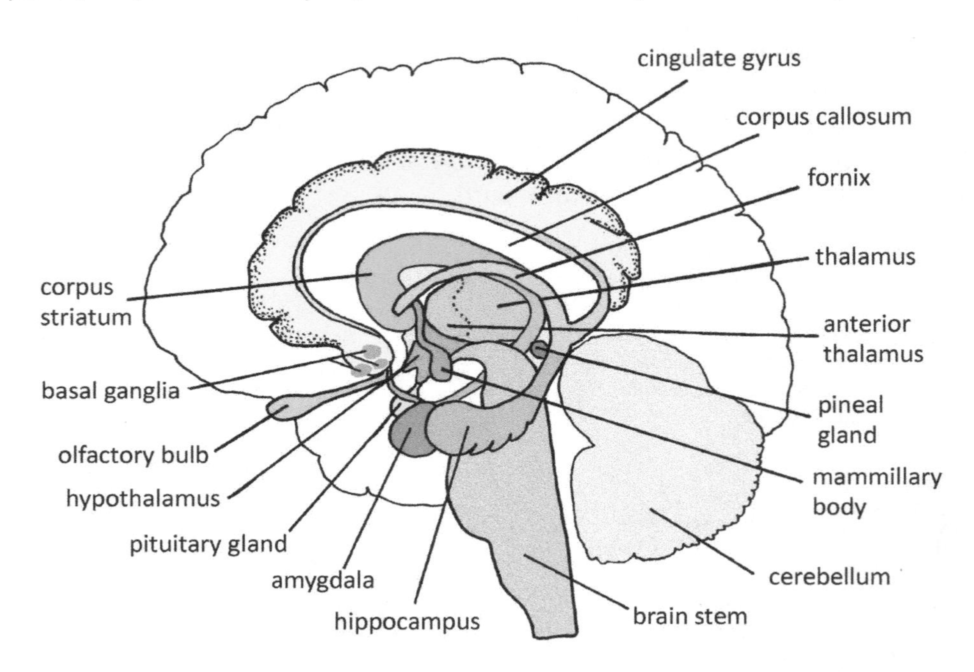


Going counterclockwise:

**CORPUS STRIATUM**: We already know that "corpus" means "body." The word "striatum" means "striped." This area has a mildly striped texture if you cut it open. The corpus striatum is connected to many other parts, though in diagrams you can't see these connections. The striatum seems to be involved in decision making, especially when it relates to movement. It also seems to play a role in motivation and in response to rewards and punishments. It is here that "conditioned" responses develop. The famous example of this is "Pavlov's dogs," where Mr. Pavlov conditioned his dogs to salivate every time they heard a bell ring because they had learned that when the bell rang they would soon be fed. Their corpus striatums connected the sound of a bell to food, and sent signals to the nerves in glands in the mouth to produce saliva. The corpus striatum is also one of the parts that controls addictions; it can make inappropriate connections to behaviors and rewards, causing the addicted person to choose patterns of behavior that are actually harmful to them.

**BASAL GANGLIA**: "Basal" means "at or near the bottom," and "ganglia" is the plural form of "ganglion" which means "tangle, knot, or clump." So the basal ganglia are little "knots" of nerve fibers that sit at the bottom of the cerebrum. These might be the strangest parts on the diagram. Their job is to stop, or inhibit, movement. They are constantly putting the brakes on your motor cortex. The reason that you can think about moving your arm or leg without actually doing it is thanks to your basal ganglia. If the basal ganglia become less active and stops sending out inhibitory chemicals to the thalamus, this will cause a person to make unnecessary motions that they cannot control (as in Tourette's syndrome). If the basal ganglia are too active, (as seems to be the case with Parkinson's Disease), the thalamus will be suppressed and won't relay signals to the motor cortex.

**OLFACTORY BULB** Although it might sound like a light bulb factory, it is really the nerve that brings in smelling sensations from the nose. Your nose hooks up to your brain very close to the place where memories are stored, the hippocampus. Some scientists think this is why you almost never forget a smell. The odor sensations are then sent to the temporal lobe where your brain determines what kind of smell it is.

**HYPOTHALAMUS:** The word "hypo" means "under," so the hypothalamus is located under the thalamus. Although this part is very small, it has lots of important jobs. It keeps your body temperature and your blood pressure constant, makes you feel hungry or thirsty, helps sort out emotions coming from the amygdala, manufactures chemicals that control growth, and sends chemical messages down to the pituitary gland telling it when to release its chemicals. The hypothalamus is quite a multi-tasker! Think of the hypothalamus as your body's balance regulator, letting other organs know when there is too much or too little of something. The scientific word for keeping the levels of things constant is ***homeostasis.***

**PITUITARY GLAND** The pituitary *(pit-TU-it-tary)* gland produces chemicals that make you grow. It also makes hormones that causes a child to mature into an adult.

**AMYGDALA:** The amygdala *(ah-MIG-dahl-ah)* is found at the end of the center for emotion, especially strong emotions such as fear and anger. It turns on the bodily functions that are related to these emotions, like increased heart rate and sweating. This emotional center just happens to be at the tip of the part that controls memory, which is possibly why our strongest memories are things we felt strongly about when we experienced them.

**HIPPOCAMPUS:** This is the Latin word for seahorse. The photograph below shows a hippocampus (with fornix attached) next to a seahorse. Pretty amazing, eh? Probably what we now call the fornix (the tail of the seahorse) was originally considered to be part of the hippocampus. The hippocampus lets you compare new experiences to old ones that you already have in your memory, and it also puts memories into longterm storage. It's sort of like a librarian, finding a place to put new books (memories) and pulling out old ones (remembering) when you need them.

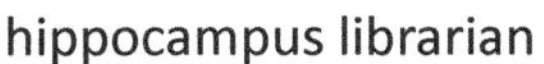
hippocampus librarian

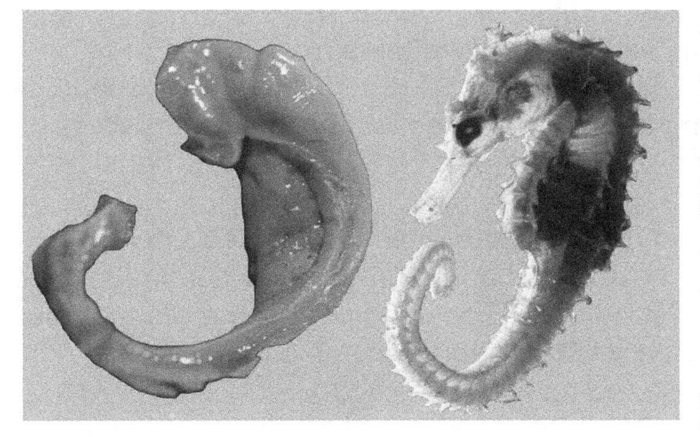

By Hippocampus_and_seahorse.JPG: Professor Laszlo Seress derivative work: Anthonyhcole (talk) - Hippocampus_and_seahorse.JPG, CC BY-SA 3.0, https://commons.wikimedia.org/w/index.php?curid=9451294

**MAMMILLARY BODY:** This tiny part is still somewhat of a mystery. What we know about it comes from cases of brain damage. It seems to play a part in memory, and might also help us remember places and travel routes.

**PINEAL GLAND:** This gland, not surprisingly, is shaped like a pine cone. The pineal gland produces a hormone called ***melatonin***, which helps to control sleeping and waking cycles. It is connected to special light-sensing cells at the back of the eye. All mammals have this gland and in some mammals it also controls seasonal cycles, such as hibernation. Recently, it has been discovered that the pineal gland also stimulates production of new bone.

**THALAMUS:** This part helps to sort out all the sensory signals coming from in from the body and then either ignore them or send them on to the appropriate lobes of the cerebrum. The thalamus also seems to be the center of ***consciousness***. Damage to the thalamus can cause a person to go into a coma (a prolonged period of unconsciousness deeper than ordinary sleep). The **ANTERIOR THALAMUS** is the front part of the thalamus. It is sometimes classified as part of the limbic system, along with the hypothalamus. It receives input from the mammillary bodies via the fornix. It is connected to episodic memory (more about this in a later chapter).

**FORNIX:** "Fornix" is Latin for "arch." This part connects the hippocampus to other brain parts.

**VENTRICLES:** These are not labeled—they are the "empty spaces" between all the parts, and are filled with cerebrospinal fluid. If you read the next section, you will learn more about them.

**ACTIVITY 4.1 "The Brain Song"**

Here is a silly song about some of the parts of the brain. You can download the audio tracks at this web address: www.ellenjmchenry.com/BrainCurriculum.

## THE BRAIN SONG

I woke up Monday morning, just like I always do;
Without my PONS to get me up, I'd sleep the whole day through.
My faithful old MEDULLA had worked all through the night,
To keep my heart and lungs working right.
Oh, my brain stem works so hard,
It does so many things I disregard (oh, how very boring)
Thinking about my CEREBELLUM is a snore,
Without it, though, my head would hit the floor.

I jumped up out of bed, then, and dressed without a fuss;
I was getting signals from my HYPOTHALAMUS.
I went out to the kitchen for cereal and a bowl,
But what a sight for my OCCIPITAL!
Pans and dishes filled the sink.
My TEMPORAL LOBE could smell the garbage stink (oh, how very awful)
Thinking about the chores that waited for me there
Was more than my poor FRONTAL LOBE could bear!

I fled that great disaster using MOTOR CORTEX nerves,
I hurdled over forks and crumbs and Friday night's hors d'oeuvres.
My FRONTAL LOBE decided to take me out the door,
But I really wish I'd seen that apple core!
As I gazed up at the ceiling,
My TEMPORAL LOBE could "hear the birdies sing" (oh, how very lovely)
Thinking I'd made my poor PARIETAL go lame,
My LIMBIC SYSTEM felt a sense of shame.

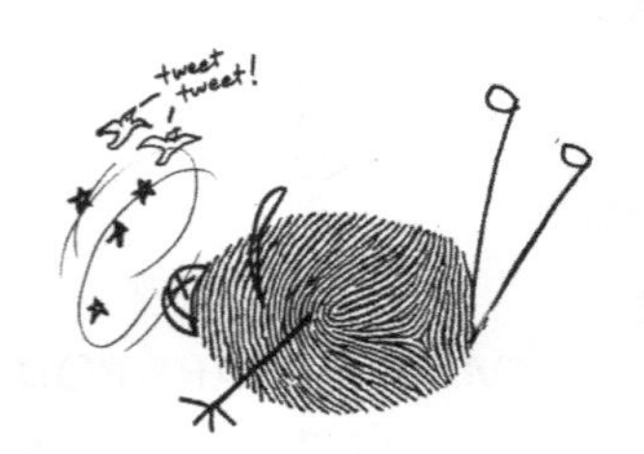


Oh, my HIPPOCAMPUS would make sure
The memory of this day would long endure (oh, how very poignant)
Thinking I'd better get some help with my hygiene,
I dialed 1-800-42-GET-IT-CLEAN!

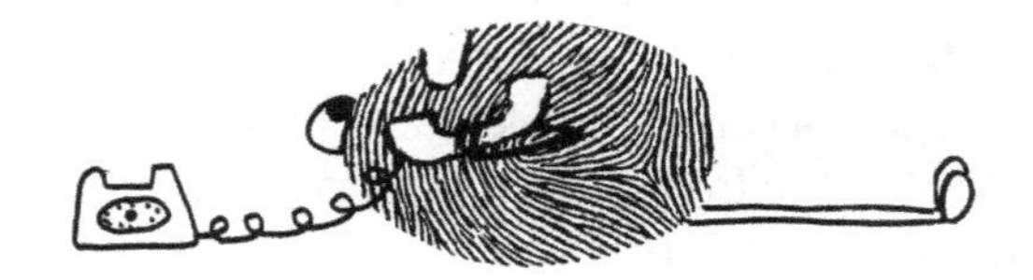

## ACTIVITY 4.2 A crossword puzzle about brain parts

ACROSS:
3) This lobe control your muscles
7) This "vaulted arch" connects the hippocampus to other brain parts.
8) This lobe keeps track of 3D space and where all your body parts are in that space
11) This lobe interprets sensory signals from your eyes
13) This part lets you know when you need food
14) This pea-sized part controls your growth
16) This brain area is in the middle and seems to "direct traffic" sorting out incoming signals
17) This lobe receives input from your ears and and nose, and also contains your speech center
18) This lobe receives signals from your sensory nerves

DOWN:
1) This C-shaped part is on the bottom of the cerebrum and connects it to the inner brain.
2) This tiny part controls strong emotions such as anger and fear.
4) The name of this lobe means "little brain." It helps with balance and coordination.
5) This is the bottom of the brain stem and controls breathing and the beating of your heart
6) This nerve "bulb" receives input from the nose.
7) This lobe is where your rational thinking occurs, such as decision making and calculations.
9) This is the part that puts memories into long term storage, and is also able to retrieve them. It is like your librarian.
10) This part of the brain stem wakes you up in the morning.
12) These are empty spaces filled with fluid.
15) This part is above the hypothalamus and seems

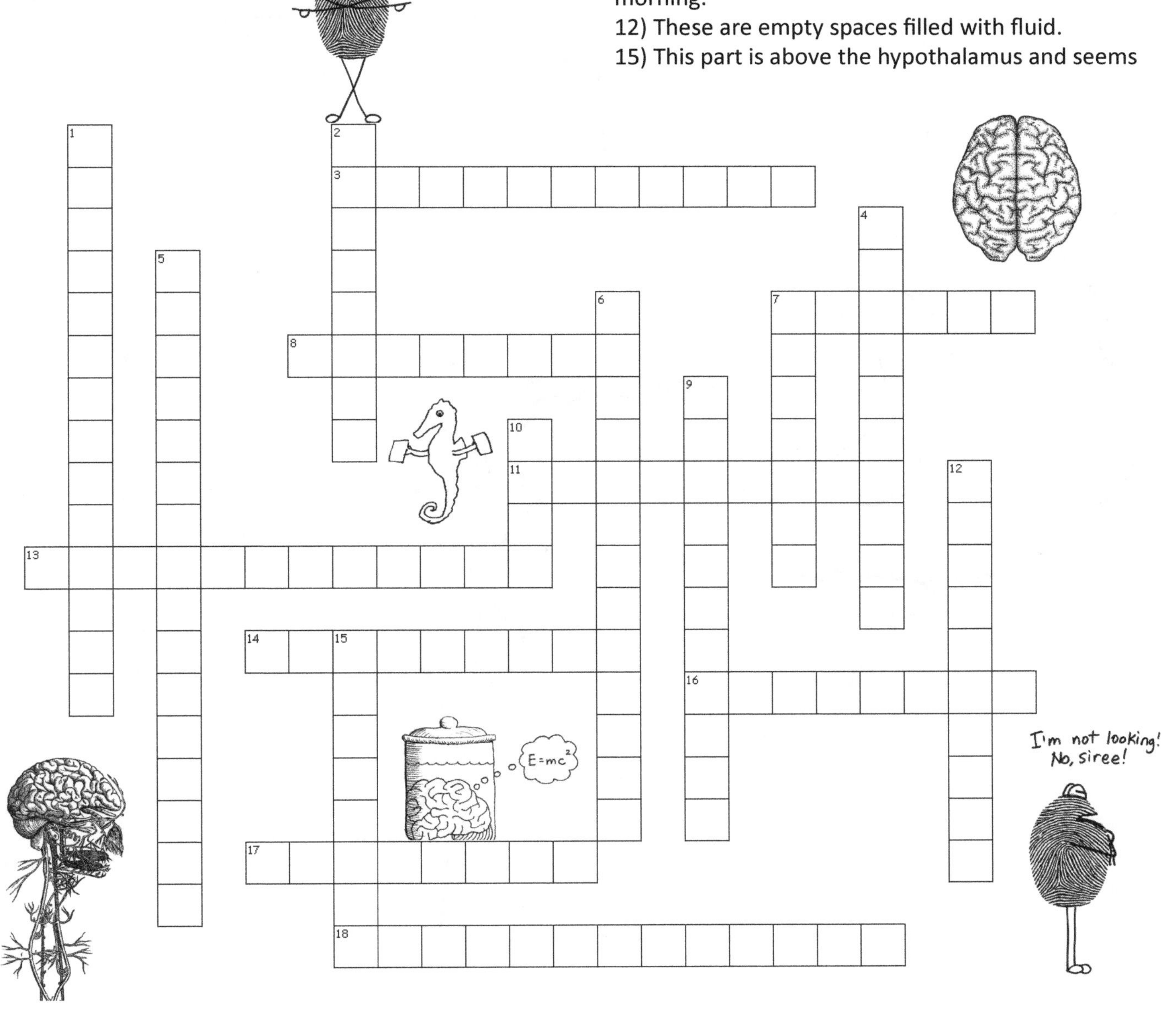

## ACTIVITY 4.3 Color-coded brain parts again

This activity is just like the one you did on page 10, except that we have added the parts deep inside the brain. Decide on a color code and make the parts of the brain match their boxes in the key. (If you run out of colors, use patterns like dots or stripes.) This diagram is not identical to either diagram in this chapter, but that's okay because brain diagrams are rarely the same. You can figure it out!

- ☐ cerebellum
- ☐ sensory cortex
- ☐ cingulate gyrus
- ☐ hippocampus
- ☐ amygdala
- ☐ mid-brain
- ☐ temporal lobe
- ☐ occipital lobe
- ☐ motor cortex
- ☐ corpus callosum
- ☐ thalamus
- ☐ pituitary
- ☐ pons
- ☐ basal ganglia
- ☐ parietal lobe
- ☐ frontal lobe
- ☐ fornix
- ☐ hypothalamus
- ☐ olfactory bulb
- ☐ medulla oblongata
- ☐ pineal gland
- ☐ mamillary body

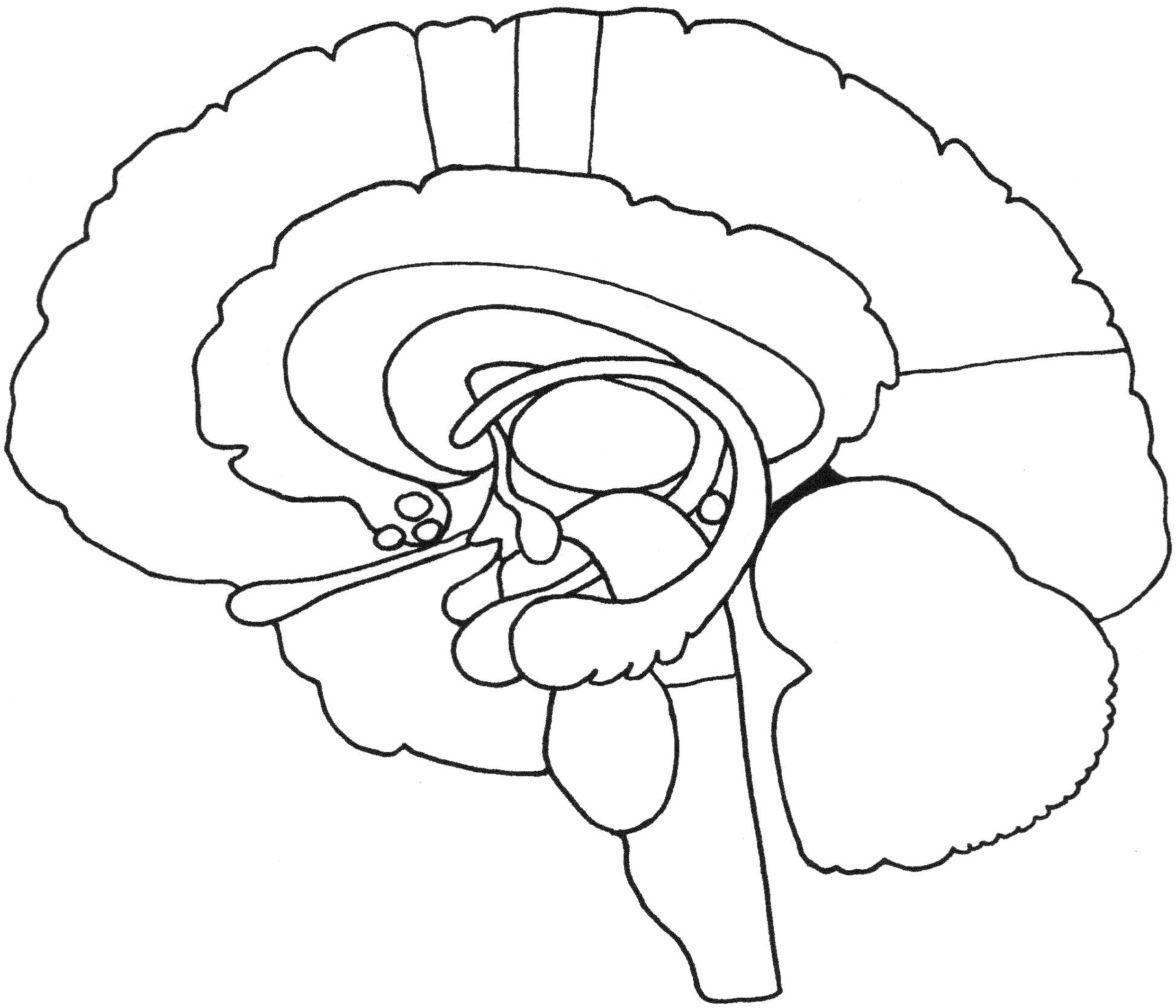

# CHAPTER 4.5

## MORE ABOUT DEEP INSIDE THE BRAIN

A feature that is found deep inside the brain and is easy to overlook is... empty space! Since empty spaces are not really brain parts, it is easy to forget about them when you are studying the brain. Yet they deserve to be mentioned because they are vitally important.

The empty spaces in the brain are called ***ventricles***. The shape of the ventricles was first discovered by pouring wax into a cow's brain. The wax filled up the ventricles, then cooled into a solid. The brain tissue was removed, leaving just the ventricle-shaped blobs of wax.

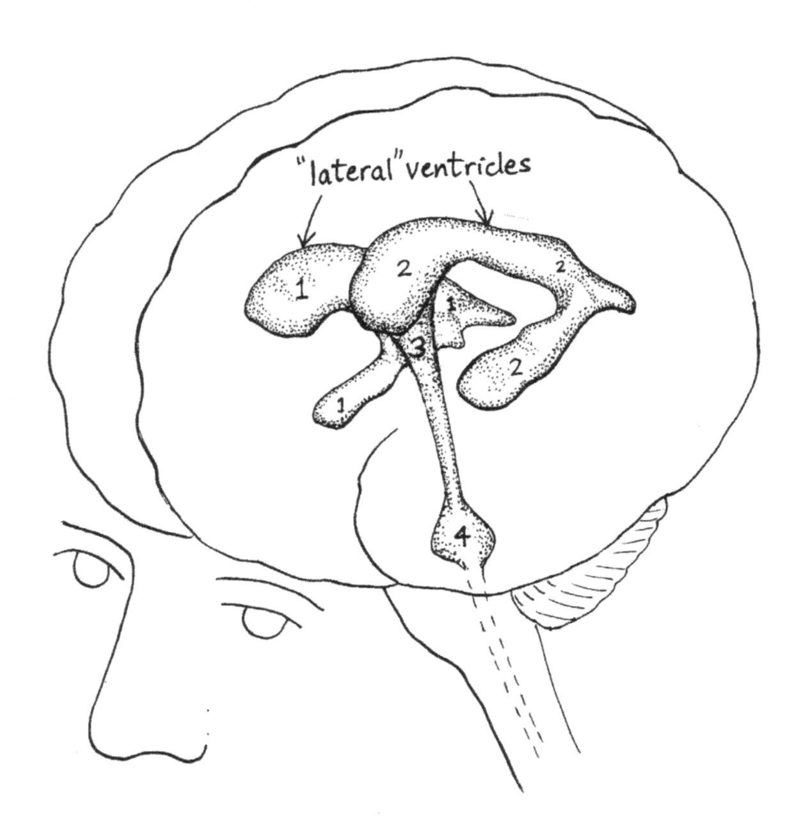


Human brains have four ventricles. Their shape is difficult to describe and just as difficult to draw. However, here is an attempt to show you what they look like (with their shapes simplified a bit). The ventricles are filled with the same fluid found between the brain and the skull: ***cerebrospinal fluid***, or ***CSF***. This fluid is produced by cells in the ventricles and consists of water, sugar (glucose), proteins, vitamins, and minerals. You may remember from a previous chapter that the brain produces about a spoonful of fluid every hour.

The fourth ventricle flows down into the spinal cord. The fluid drains down through the center of the spinal cord, and then out the bottom to be reabsorbed and recycled by other body tissues. If, for some reason, the fluid cannot drain properly, the result can be that the ventricles become over-filled with fluid and start to expand, putting pressure on the brain. To correct this problem, surgeons can install a plastic drainage tube (called ***shunt***) that allows the extra fluid to drain down into the stomach.

Spaces inside the brain are filled with either brain cells or cerebrospinal fluid, never air or blood. If someone has surgery to remove part of their brain, that empty space will fill up with cerebrospinal fluid. If the brain begins to shrink, due to disease or aging, the brain fills that space with fluid. Doctors can look at MRI scans of the ventricles to help them diagnose diseases such as schizophrenia or Alzheimer's. If they see large ventricles, that means the brain has gotten smaller.

We do not know for sure why there are ventricles deep inside the brain, but here are some good guesses. First, they help provide protection for the interior parts. Second, they sort of "float" the brain so it doesn't press down so heavily on the top of your spine and neck. (If you have ever picked someone up while in a swimming pool, you know that objects seem far less heavy when they are floating in water.) Third, the extra fluid may be necessary to nourish the inner brain parts that do such important jobs, such as regulating your body temperature and your appetite. The fluid delivers nutrients to the brain cells and removes wastes. Your brain needs those ventricles!

## ACTIVITY 4.5 Color two more PET scans

Color these PET scans using the number code. You will be able to compare the scans and see how Alzheimer's can cause loss of brain tissue, thus creating larger ventricles.

1 = yellow or light green   2 = blue   3 = purple   V = leave blank (ventricle)

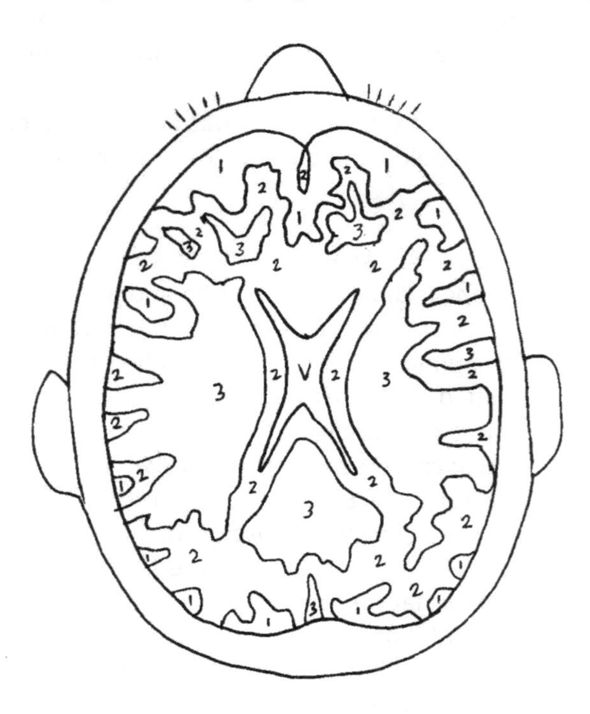

Normal patient

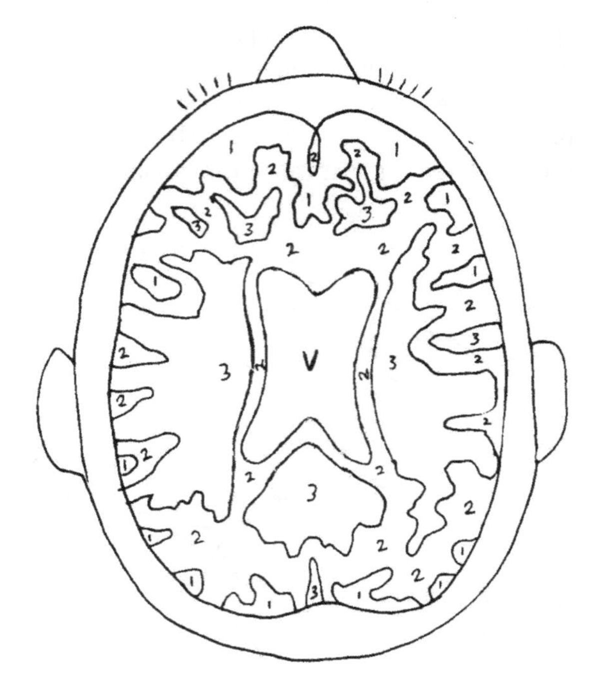

Patient with Alzheimer's

## ACTIVITY 4.6 More about ventricles

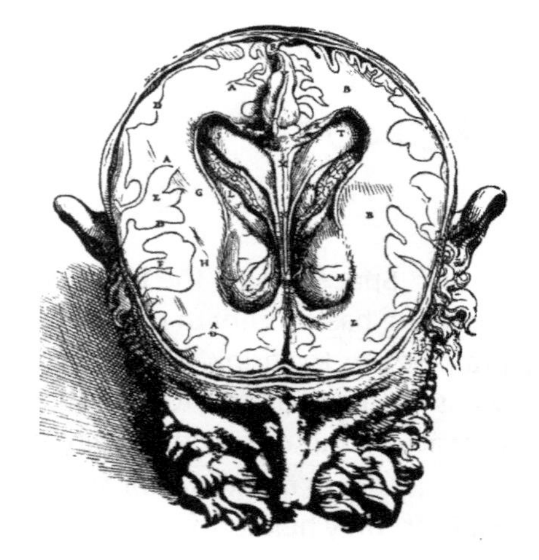

To get a better idea of what the ventricles look like, you need to see them from different angles and with brain parts in and around them. You can use Google or some other image search engine and the key words "brain ventricles" to look at a variety of ventricle illustrations. You will notice how different they look in each illustration.

You may turn up one of Vesalius' illustrations from the 1500s, such as the one shown here. (Remember Vesalius from way back in chapter 1?) In case you are wondering, he probably did not actually see a real head cut this way. He liked to be creative with his drawings and thought it would be humorous to make this one look like a living head.

There are also some videos abut ventricles posted on the YouTube channel (**www.YouTube.com/TheBasementWorkshop**. Ventricles may seem an unlikely topic for someone to make a video, but actually there are quite a few good videos to choose from. Several of the videos use accurate 3D models of these spaces, so you see how they are connected and how they relate to the brain lobes around them. Happy ventricle viewing!

# CHAPTER 5

## BRAIN CELLS

The basic building blocks of your body are ***cells***. There are different kinds of cells, designed to do particular jobs. For example, the cells in your skin were designed to cover and protect you, the cells in your muscles were designed to contract, the cells in the back of your eyes were designed to sense light, the cells that line your sinuses were designed to produce mucus, and red blood cells were designed to carry oxygen. The size and shape of the cell depends on its job.

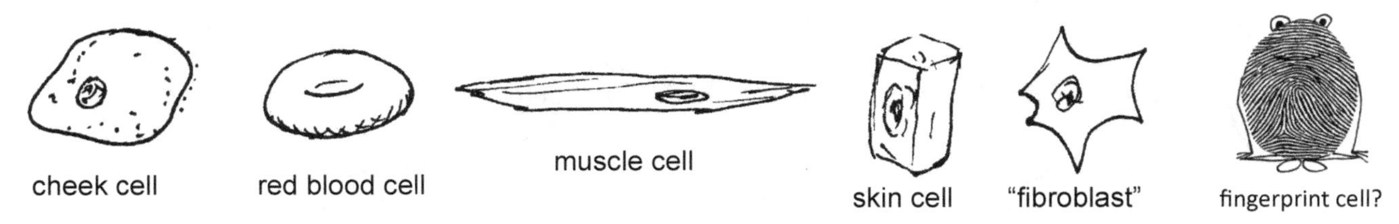


What do brain cells need to do? They must be able to generate chemical and electrical signals. The electrical signals in your brain cells are not like the electrical current found in the wires of your house. Some cells found in other animals—electric eels, for example—are able to generate large electrical currents, but fortunately your brain cells can't do this.

There are actually two types of cells in the brain: ***neurons*** and ***glial*** *(GLEE-ul)* cells. Neurons get all the attention because they have the exciting job of producing the electrical signals. However, there are ten times more glial cells in the brain than there are neurons, so they must be very important. We aren't going to forget those glial cells and will come back to them in a minute, but first, let's look at neurons.

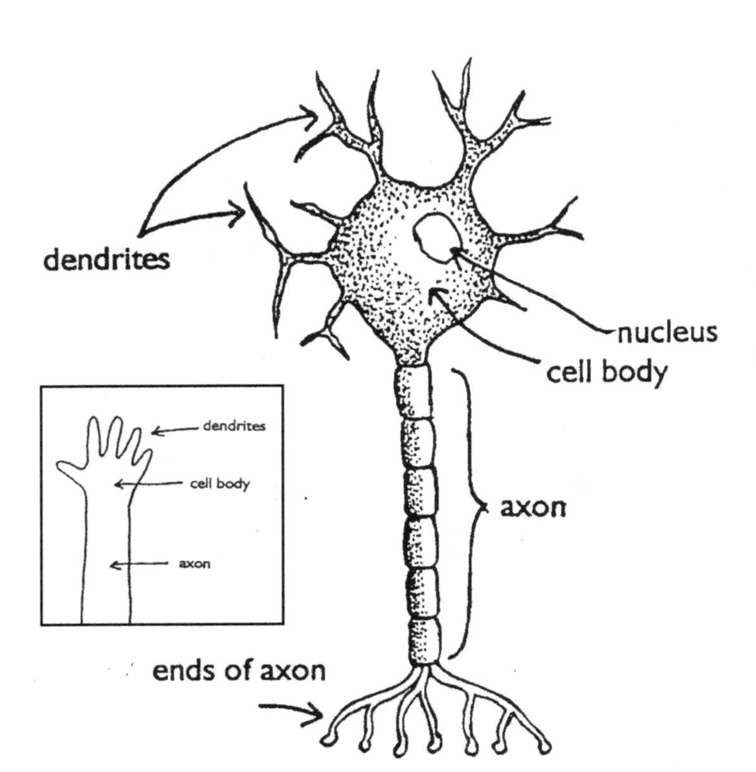


If your forearm was a neuron, the fingers would be dendrites.

Neurons are strange-looking cells. They have a large area, called the ***cell body*** with lots of little branches coming out of it. These little branches are called ***dendrites***, which is Greek for "tree branches." If you made a neuron model with your hand, your fingers would be the dendrites.

The long skinny part is called the ***axon***. The end of the axon branches off, but they are not called dendrites, they are just the ends of the axon. Notice that at the ends of the axon are little knobs (the ***terminal knobs***). We'll find out what they do on the next page.

The neuron's electrical signal goes only one way. It starts in the cell body, travels down the axon, and stops at the ends of the axon. It never goes the other way. The branching dendrites are like antennae, or receivers, that "listen" for signals from other cells. The knobs at the ends of the axon are like transmitters that send the signal out to other cells.

This neuron is actually the type that is found outside of the brain, the kind that makes the nerves found throughout the rest of your body. Brain neurons don't have those marshmallow-looking things around their axons. More about this on the next page.

Inside the main body of the cell are little parts called ***organelles***. The organelles you find inside a neuron are the same organelles you find in all the other cells of the body. The nucleus is the place where the DNA (deoxyribonucleic acid) is located. The DNA provides all the instructions for the cell so it knows how to do its job. The mitochondria's job is to produce energy for the cell. Many of the other small parts are like little factories or warehouse that make, store, and ship complex molecules that the cell needs.

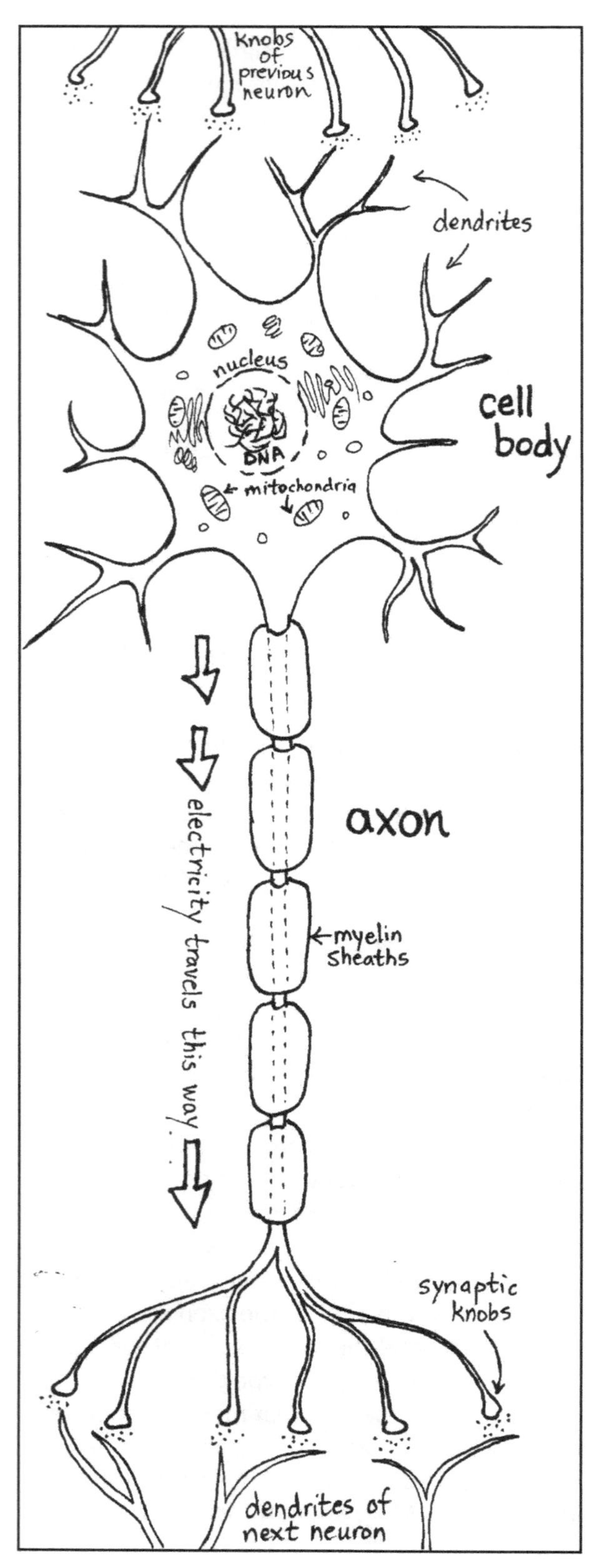


The pads along the axon—looking like a string of sausages or marshmallows—make the ***myelin sheath***. ("Sheath" just means "cover.") They act like the rubber insulation around electrical wires that keeps the electricity inside the wires. Myelin *(MY-ell-in)* is made mostly of fat, with a little protein mixed in, and is a very good insulator. The myelin allows the electrical signals to travel at speeds up to 270 miles per hour (435 km/hr). Each "sausage" (or marshmallow) is actually a living cell, albeit an extremely weird cell that is flatter than a pancake and is all rolled up. These cells are called ***Schwann cells*** (after Mr. Schwann, who discovered them). Schwann cells are found around nerves in the body. In the brain, the myelin sheaths are produced by a special kind of helper cell you'll meet in chapter 4.5.

The little knobs at the ends of the axons have a special job. They must relay the electrical signal to the next neuron down the line. However, it's not a simple task, because neurons don't actually touch each other. There is a microscopic gap between them, called the ***synapse*** *(SIN-aps)*. The electrical signal cannot jump across this gap. The knobs must make chemicals that act as little ferry boats, crossing the gap. When these little ferry boats reach the far shore (the dendrites of the next neuron) they deliver their message: "Hey, cell! Start an electrical signal immediately!" These chemicals are called ***neurotransmitters***. After they cross the gap, they must be recycled. The neurotransmitters might be dissolved or they might be gathered up and put back into the terminal knobs. This process happens very quickly so that you can make the same motion repeatedly.

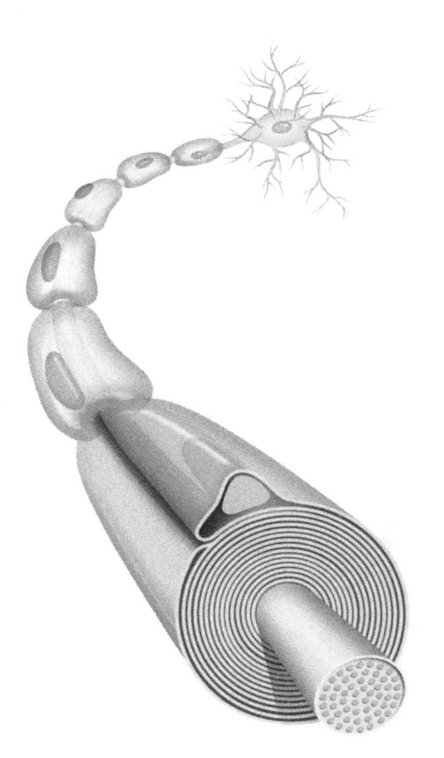

Scwann cells are blue. Yellow is axon. The pink thing is the Schwann cell's nucleus. Look how many time the Schwann cell is wrapped around the axon!

As you read the following paragraph, you might want to put your finger on the first neuron (on the left) and move your finger along with the described actions.

The first neuron starts an electrical signal. It travels down the axon until it reaches the knobs. The knobs make chemicals that go across the gap. The chemicals deliver their message to the next neuron and it begins to fire off an electrical signal. The electricity travels down that axon until it reaches the knobs, then stops. The knobs make chemicals that cross the gap and deliver the message to the next neuron. The next neuron fires off an electrical signal. The electricity travels down the axon until it reaches the knobs. The knobs make chemicals. And it continues on this way, alternating between electricity and chemicals, until the signal finally reaches its destination, wherever that is. The destination might be a muscle cell, in which case the electrical signal would cause the muscle cell to contract. Some signals stay inside the brain. (In which case, we wouldn't see Schwann cells.)

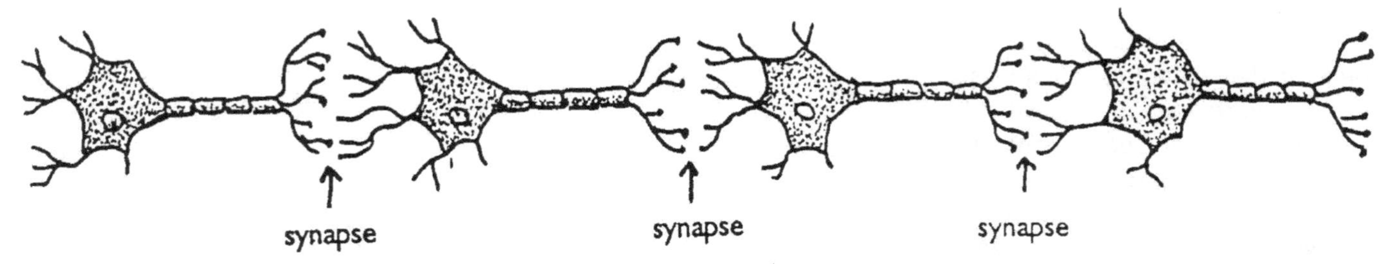


Now, what about that other kind of cell, the one everyone tends to forget about? Those poor unsung heroes are the ***glial cells***. The word *glial (GLEE-ul)* is related to the word "glue." Glial cells act a bit like glue in the brain, holding all the neurons in place. You don't want those delicate axons and synapses getting bumped out of place! For every one neuron there are at least ten glial cells, possibly as many as fifty.

Glial cells are the caretakers of neurons. They feed them, clean up after them, protect them, and repair them if they get damaged. All the neurons need to do is send their signals. The neurons in this diagram are yellow.

There are several kinds of glial cells. The green cells in this diagram are called ***astrocytes.*** ("Astro" means "star.") This diagram only shows a few astrocytes, just to keep the picture simple. In reality, many astrocytes surround the neurons. Astrocytes protect the synapses so they don't get bumped out of place. The process of learning requires pathways to be formed from neuron to neuron, and these paths included a route through specific synapses, so they must stay lined up. You can see how the astrocyte "hugs" a blood vessel (the red tube). The astrocyte absorbs oxygen and food from the blood, then passes it along to the neurons.

The blue cells are called ***oligodendrocytes*** and are the equivalent of Schwann cells because they form rolled covers around axons. The round part of these cells is the cell body. One cell body has a few ("oligo" means "few") dendrites going out from it, and the end of each dendrites forms a sheath that will wraps around an axon.

The dark red cells are ***microglia*** and are the infection fighters inside the brain. The pink cells at the top produce cerebrospinal fluid and are therefore found lining the walls of the ventricles.

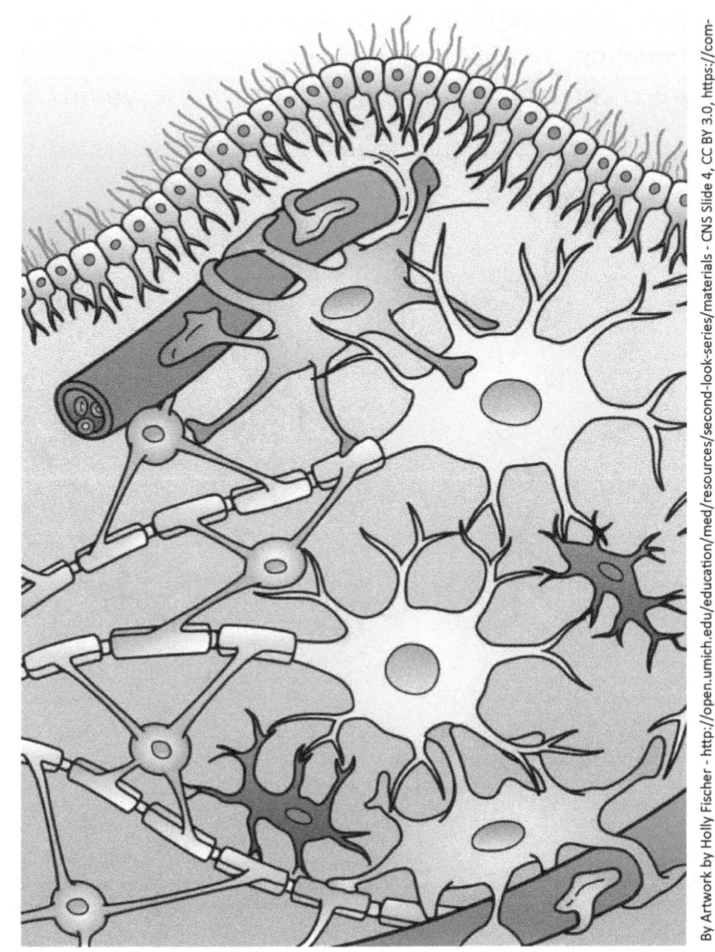

Here is another diagram showing the same thing. In this drawing, the artist has not shown us the astrocyte touching a blood vessel, but we do see the astrocytes connecting with the dendrites of the ***ependymal*** cells (eh-PEND-i-mal). Touching the astrocytes allows the ependymal cells to communicate with the other brain cells. This is import when neurons or other glial cells become damaged. In 1999, it was discovered that an ependymal cell can change to become one of the other cells types. This is very helpful when the brain is trying to recover from a stroke, one of the brain injuries we mentioned in chapter 1. In a stroke, one of the brain's tiny blood vessels becomes blocked and the brain cells die due to lack of oxygen.

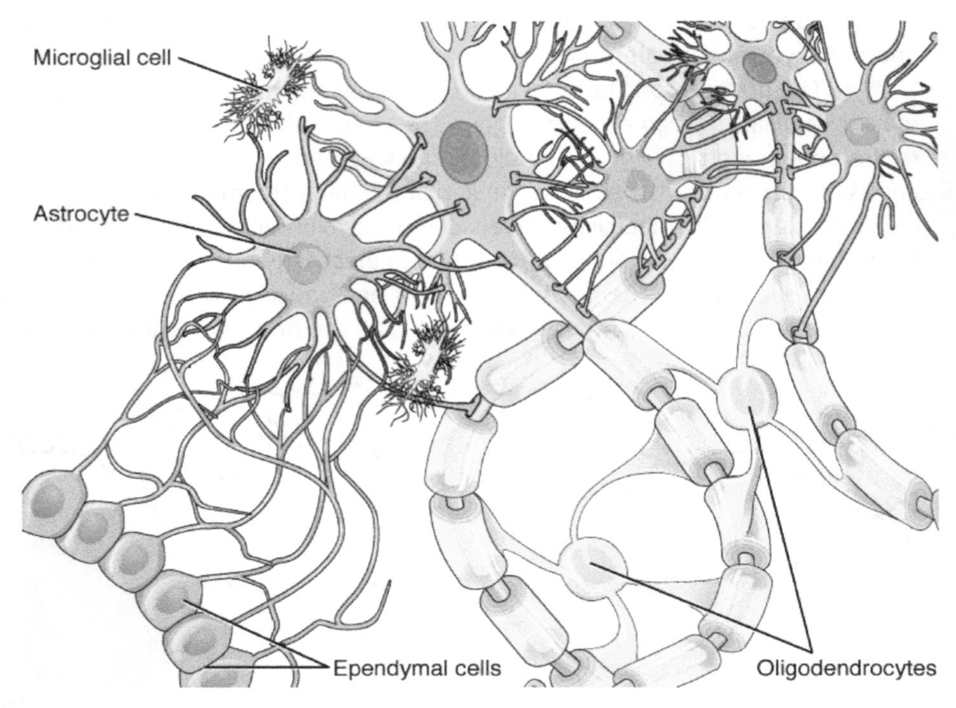




Now let's go back to talking about neurons for a minute. We know the basic structure of all neurons: they have dendrites, cell bodies, and axons. The exact shape of neurons can vary quite a bit, however, depending on their location and their job. They can have long dendrites and short axons, or short dendrites and long axons. They can be almost symmetric (balanced) or they can look lopsided.

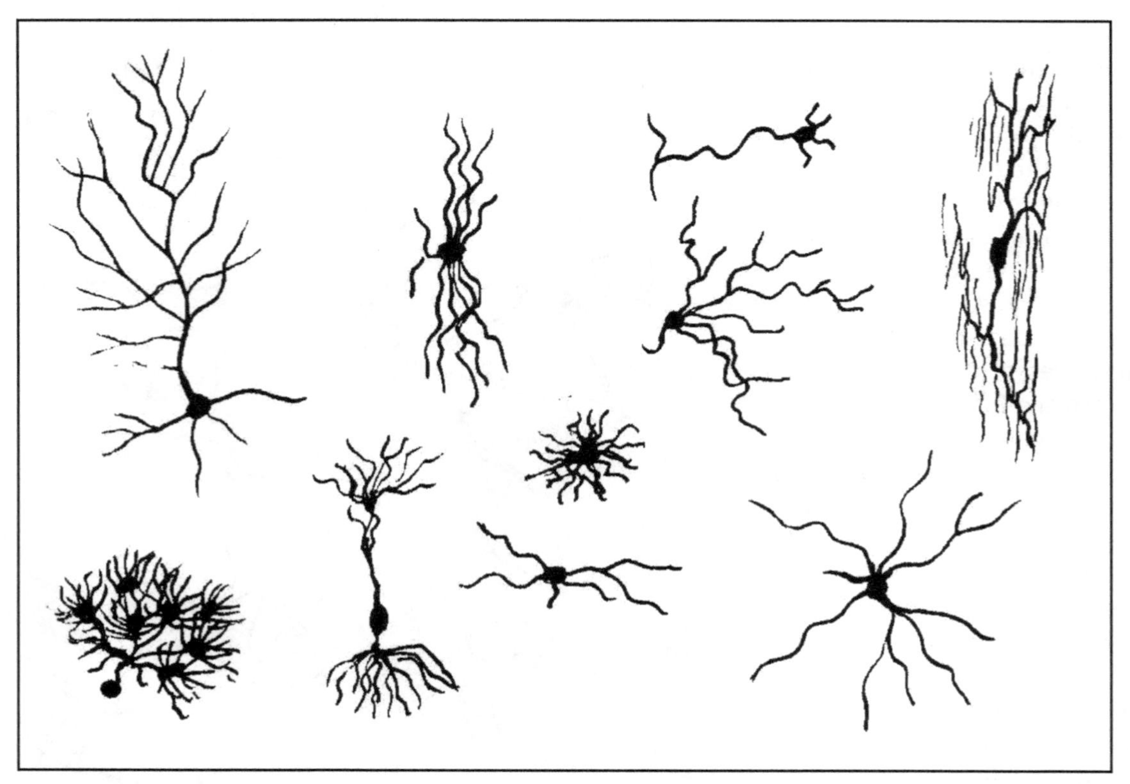

Can you spot the axon in all of these neurons? (There's no answer key, so don't worry if you guess wrong.) In several of them, the dendrites are very long and the axon is the only short one. Can you find a neuron with a very long axon that splits in two at the end? The extremely bushy one is found in the cerebellum and is called a ***Purkinje cell*** (*pur-KIN-gee),* named after the discoverer.

## A FAMOUS BRAIN SCIENTIST

Santiago Ramón y Cajal *(Ra-mone ee ka-HALL)* was a Spanish neuroscientist who lived from 1852 to 1934. As a child, Cajal was trouble-maker. At age eleven he actually went to jail for a short time because he destroyed a neighbor's garden gate with a homemade cannon. Though he did not do well in school, he was an excellent gymnast and learned to draw and paint quite well. Even though his father was an anatomy teacher at a medical school, he doubted that his son would ever have an academic career so he apprenticed him to a shoemaker.

However, one summer, young Santiago went with his father to a graveyard to pick up bones that could be studied by his father's anatomy students. Santiago was allowed to make sketches of the bones, and these drawings were so good that it was obvious to everyone that he should be allowed to try going to medical school. He did well, and after serving as an army doctor for a few years, he became a professor at several major universities.

Cajal became an expert at examining tissues under a microscope and was especially interested in looking at brain tissue. He learned about a staining technique that would stain only neurons and leave the background blank. This was just what he needed to study the patterns made by neurons. He discovered the synapse, and suggested that connection between cells' dendrites was mechanical process necessary for learning. In 1906 he was awarded the Nobel Prize in Medicine.

Here are some of Cajal's most famous drawings:

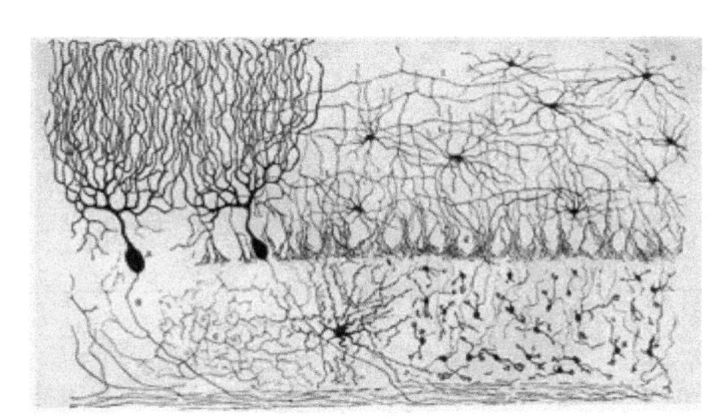

Cells from the cerebellum of a chicken.

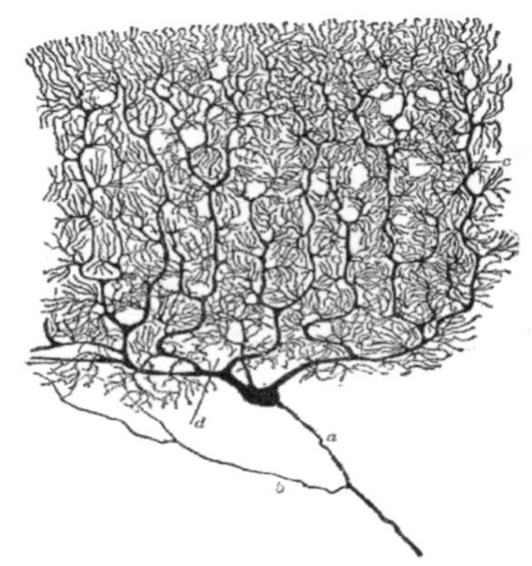

Purkinje cells from a human brain.

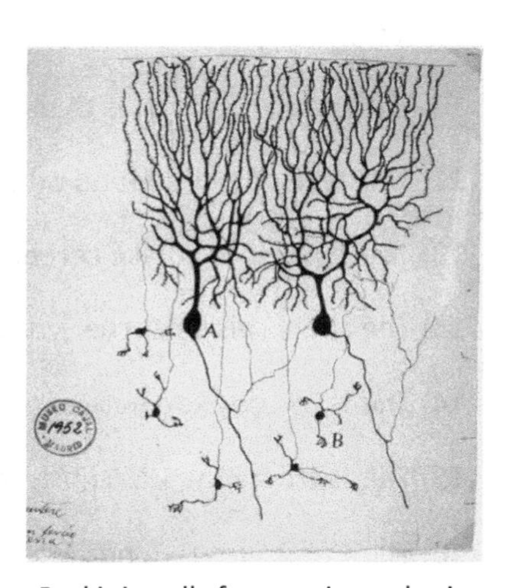

Purkinje cells from a pigeon brain.

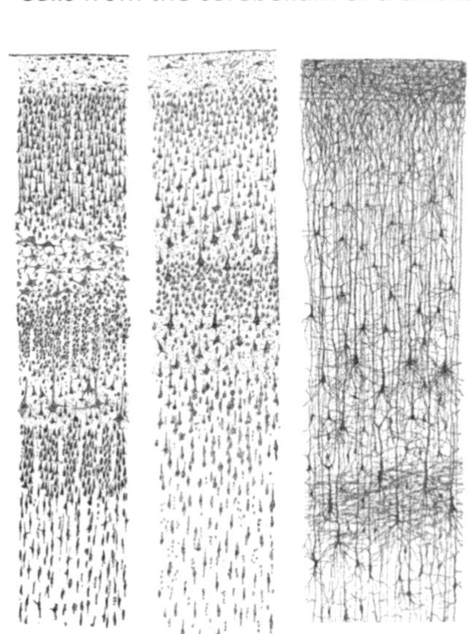

Cells from a human sensory cortex.

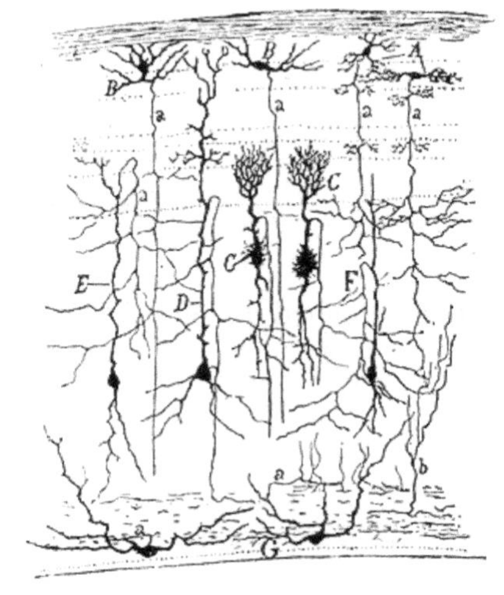

Cells from the optic lobe of a bird brain.

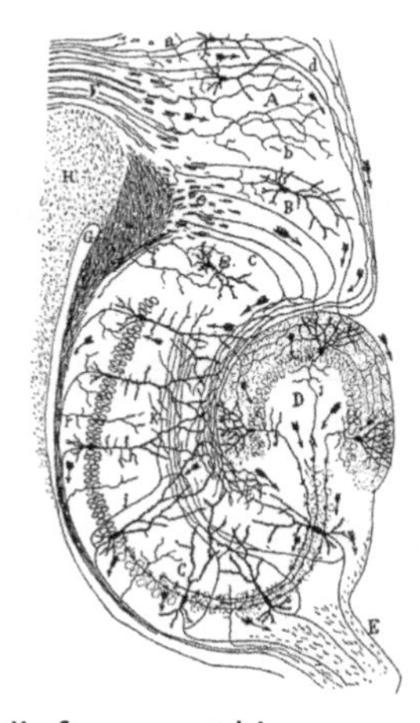

Cells from a rat hippocampus.

## ACTIVITY 5.1 Neuron videos

Don't forget to check out the neuron videos on the Brain playlist.

## ACTIVITY 5.2 Vocabulary and riddles

Read each clue and write the correct word, with one letter on each line. After you are finished, used the numbers under the lines to write the correct letters to form the answer to the silly riddles.

1) The type of cell that sends electrical signals: ___ ___ ___ ___ ___ ___
9

2) This part of a brain cell contains the nucleus and organelles: ___ ___ ___ ___ ___ ___ ___ ___
6

3) This type of cell wraps around the axons of nerve cells found outside the brain: ___ ___ ___ ___ ___ ___ ___
17 21

4) These branch off the cell body of a neuron. ___ ___ ___ ___ ___ ___ ___ ___ ___
18

5) This fatty substance is what provides the insulation around axons: ___ ___ ___ ___ ___ ___
1

6) These cells are the infection-fighting cells of the brain: ___ ___ ___ ___ ___ ___ ___ ___ ___
16

7) This type of cell attaches to both capillaries and neurons: ___ ___ ___ ___ ___ ___ ___ ___ ___
3

8) This type of cell insulates brain axons: ___ ___ ___ ___ ___ ___ ___ ___ ___ ___ ___ ___ ___ ___ ___
15 7

9) This is a very small blood vessel: ___ ___ ___ ___ ___ ___ ___ ___ ___
11 12

10) The type of brain cells that outnumbers neurons 10 to 1: ___ ___ ___ ___ ___
13

11) This type of cell is found only in the cerebellum: ___ ___ ___ ___ ___ ___ ___ ___
14

12) This type of cell make cerebrospinal fluid: ___ ___ ___ ___ ___ ___ ___ ___ ___
2

13) The chemicals cross the synapse: ___ ___ ___ ___ ___ ___ ___ ___ ___ ___ ___ ___ ___ ___ ___ ___ ___
4 10

14) This is the tiny gap between neurons: ___ ___ ___ ___ ___ ___ ___
19

15) This is where the DNA (instructions) are found in a cell: ___ ___ ___ ___ ___ ___ ___
5

16) The Greek word "dendrite" means: ___ ___ ___ ___ ___ ___
8

17) This part of a neuron is very long and skinny, and wrapped by insulation: ___ ___ ___ ___
20

RIDDLE #1 What has a head, but no brain? ___ ___ ___ ___ ___ ___ ___
1 2 3 4 5 6 7

RIDDLE #2 What happened to the man who had his left hemisphere removed?

___ ___ ' ___ ___ ___ ___ ___ ___ ___ ___ ___ ___ ___ ___ !
8 9 10 11 12 13 14 15 16 17 18 19 20 21

## ACTIVITY 5.3 Color these brain cells

You've seen several images of brain tissue. Now it's your turn to choose the colors. Put the color into the square, then color the appropriate areas on the diagram. Cerebrospinal fluid can be white if you want to do less coloring. Also, draw some cell parts inside that neuron's cell body. (Make it look like the other cell bodies.)

- ☐ astrocytes
- ☐ oligodendrocytes
- ☐ neurons
- ☐ cerebrospinal fluid
- ☐ microglia
- ☐ ependymal cells
- ☐ capillary
- ☐ red blood cell

**ACTIVITY 5.4 Optional: Draw some brain cell cartoons**

Science doesn't always have to be serious. Here is a cartoon called "Glial Cell Appreciation Day" (for those heroes who do all the work of keeping neurons alive and well).

If you like to draw cartoons, you can use the space below to make your own silly brain cells. Brain and nerve cells have such strange shapes that you don't have to worry about making them look "right." Any way you draw them will be just fine! (Your cartoons do not need to look like these.)

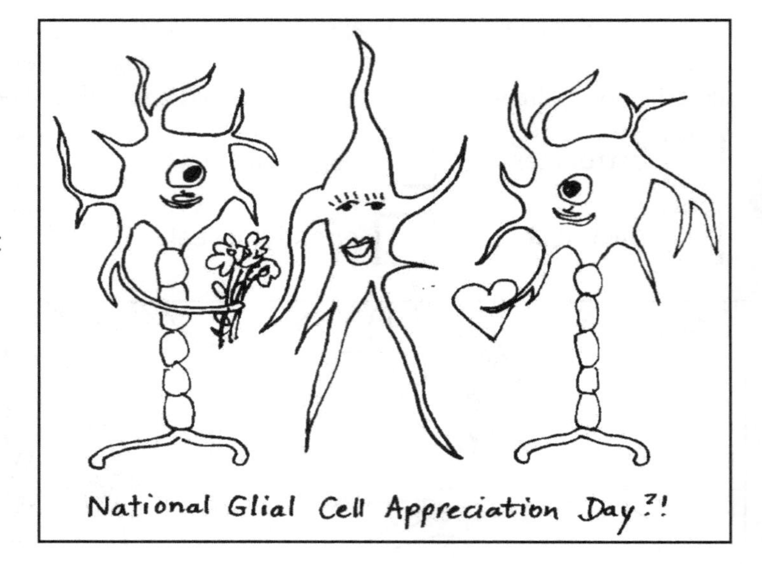
National Glial Cell Appreciation Day?!

# CHAPTER 5.5

## MORE ABOUT NEURONS: HOW THEY MAKE ELECTRICAL SIGNALS

The electrical signal made by a neuron isn't like the electricity that flows through the wires of your house. The neuron's source of positive and negative electrical charges is atoms of sodium and potassium. These atoms come from the food and drink that you take in. Both sodium and potassium are very common atoms and most types of food have plenty of both. The atoms end up all through your body, but some will find themselves near a neuron.

If you could put a tiny electrode into a neuron and measure the electrical charges both inside and outside the cell, you would find that the inside of the cell had a more negative charge than the outside. This difference is called the ***potential*** because it's this difference that gives the neuron its *potential* to work. This is the normal or "resting" state of neurons: the "***resting potential***." Strange as it may seem, unbalanced (more negative inside than outside) is "normal" for a neuron.

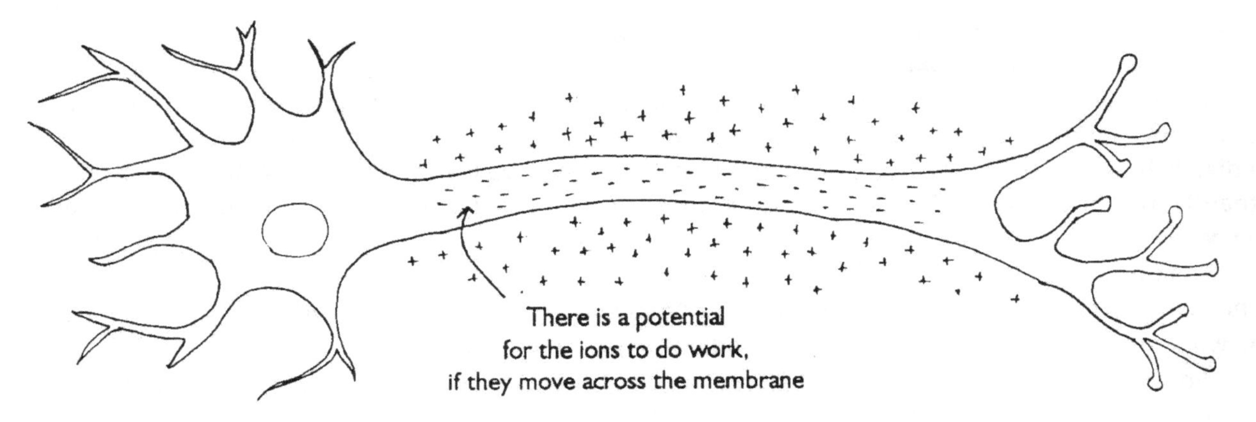

There is a potential
for the ions to do work,
if they move across the membrane


Both sodium and potassium atoms have a positive electrical charge (+), which makes them ***ions***. An ion is any atom, or piece of atom, that has an imbalance of electrons and protons and therefore carries a positive or negative charge. If the atom has more protons it will have a positive charge. If it has more electrons, it will have a negative charge. The membrane of the axon contains tiny pumps that have the ability to pump ions in and out of the cell. It pumps three sodium ions out for every two potassium ions it lets in. Two in, three out. Two in, three out. You can see that the more it pumps, the more positive the outside becomes, compared to the inside. (There would be other types of ions floating around, too, but we are only showing the sodium and potassium to keep things simple. Na stands for sodium because years ago it was called "<u>na</u>trium." Potassium used to be called "<u>k</u>alium.")

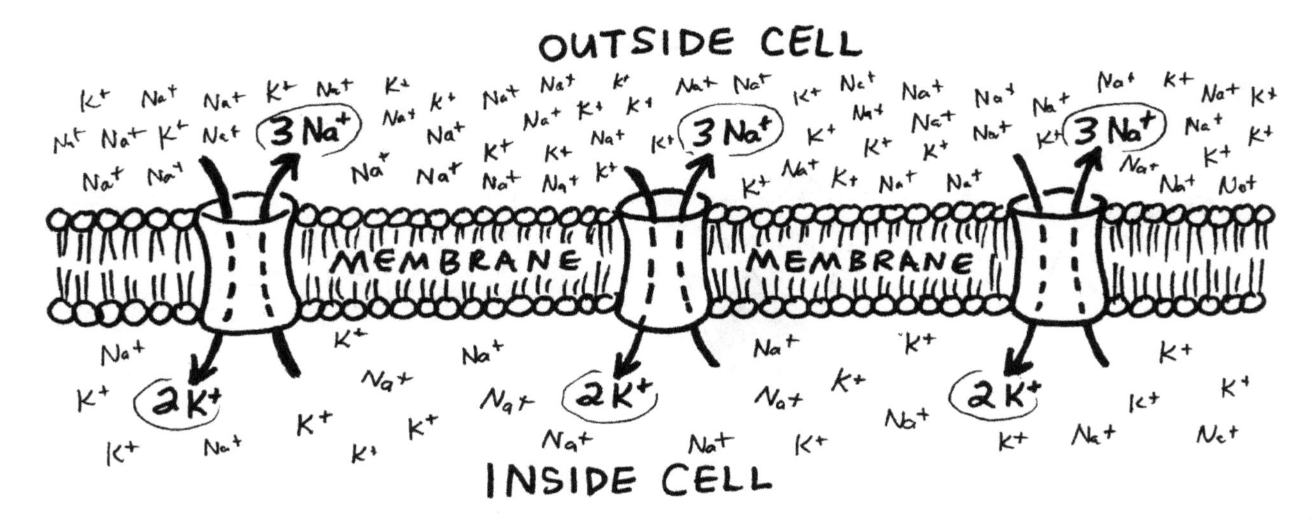

The outer membrane of the axon has tiny "gates" all along it. They are usually closed to sodium, keeping sodium on the outside (1). When a nerve impulse comes along, the number of ions outside the gates becomes so great that the gates burst open (2), allowing sodium ions to come flooding in, until there are equal numbers of sodium ions inside and outside the cell. The gates snap back to their closed position (3), and then the cell has to begin pumping all over again (4). Two K's in, three Na's out. Two K's in, three Na's out. Eventually it manages to get things back to "normal" with more (+) charges outside than inside.

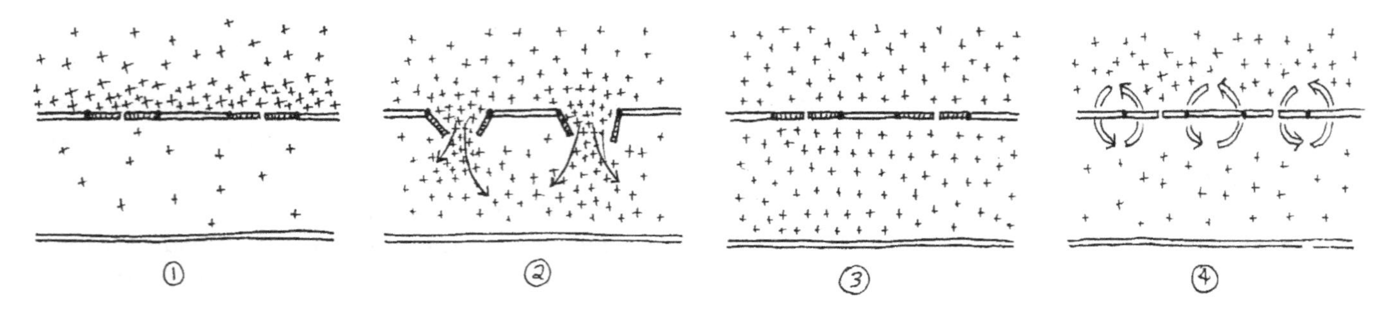


The gates are either open or closed—they can't be partially open. It's an "all-or-none" situation. The signal strength that is required to open the gates is called the ***threshold***. If the signal is below the threshold strength, the gates will remain closed. As soon as the signal hits threshold strength, the first gates snap open. Even if the signal is much stronger than threshold level, the result is the same: the first gates snap open.

This opening and closing of gates starts at the dendrites, then moves into the axon. Section by section, the gates open and close all the way down the axon. As soon as the gates slam shut, that section of the axon immediately begins resetting itself by pumping two in, three out, two in, three out. (And all of this happens in less than 1/1000 of a second!) This continuous action of gates opening and closing down the axon is called the ***action potential.***

The opening and shutting of the gates down the axon is almost like a weird sort of domino rally, with each domino being a section of the axon. But unlike the dominoes, which take a lot of work to reset, the neuron does all the work of resetting automatically, with its sodium-potassium pumps.

Now we are going to add an extra twist to the story of how the signal goes down the axon. To make the journey even faster for the electrical signal, it takes "short cuts." Wherever there is a section of myelin padding, (those marshmallow or sausage looking things), the signal can sort of skip right over that section and jump right to the next one. The tiny gaps between the Schwann cells (or oligodendrocytes in the brain) are called the ***nodes of Ranvier***. (Guess who discovered them... yep, Mr. Ranvier.)

Videos of this can be very helpful. There are some videos posted on the Brain Curriculum playlist on the YouTube channel.

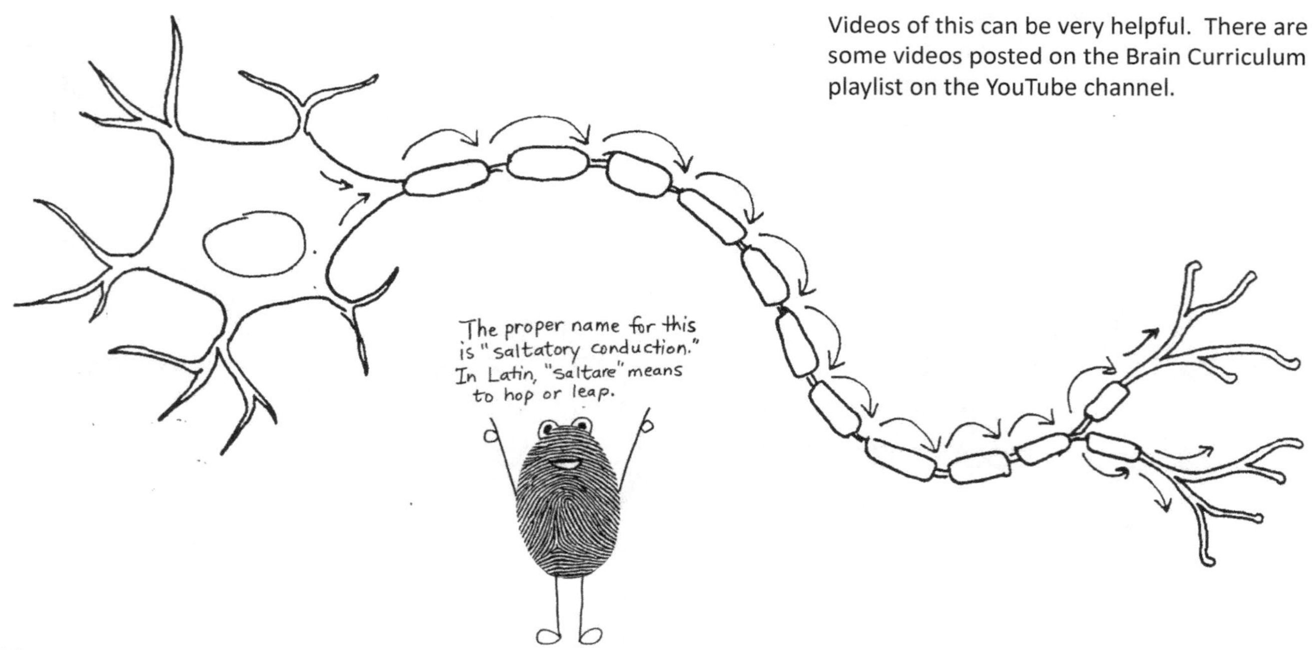

The proper name for this is "saltatory conduction." In Latin, "saltare" means to hop or leap.

So the signal skips along from node to node, without having to go the complete distance in between. To go back to our domino analogy, it would be like having an extra long spacing bar representing the myelin sheath area. The signal is strong enough to push the bar, which immediately sets off the next section, without having a bunch of dominoes in between.

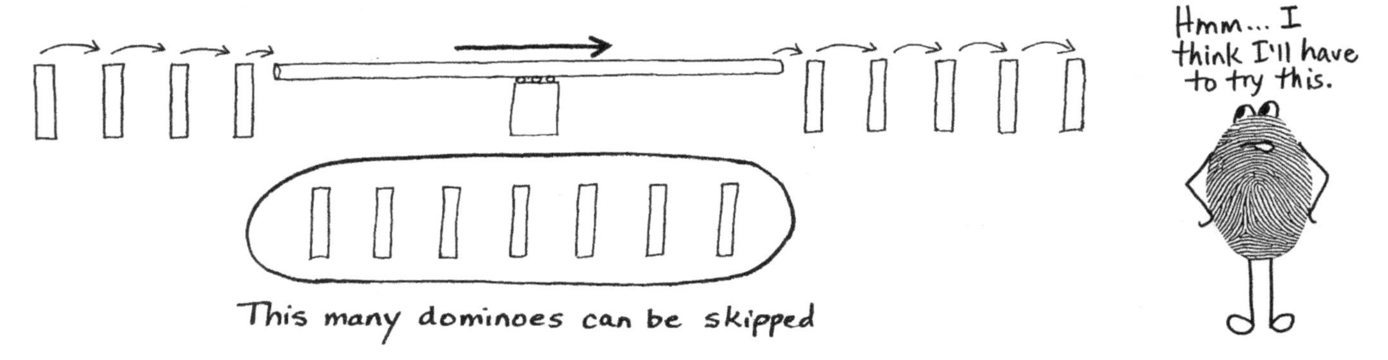


Dominoes seem to fall very quickly, so we might not think this extra speed is very important. But to the neuron, it makes a big difference. Signals such as reflexes can't be just fast, they have to be lightning fast. Your reflexes need to do things like pull your hand back from the stove in a split second.

Now we need to look at what happens to the nerve impulse when it reaches the synapse. As we've already learned, when the electrical signal reaches the end of the axon, it comes to a screeching halt. It cannot cross the gap. The electrical signal must be translated into a chemical signal that can cross the synapse.

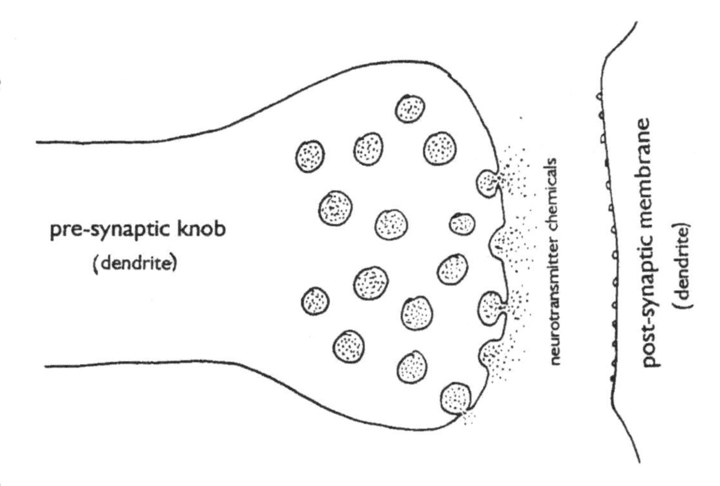


The ends of the axons look like little knobs. They are called the ***pre-synaptic knobs***, since they are they come before (pre) the synapse. The electrical signal allows calcium ions to flow into the cell, causing little bubbles (***vesicles***) full of chemicals to "fuse" with (sort of melt into) the outside wall of the knob, thus spilling their contents into the gap. There could be tens of thousands of vesicles in each synaptic knob.

These neurotransmitter chemicals go across the synapse and touch the membrane of the dendrite on the other side. There are special places on the dendrite's membrane called ***receptor sites*** where these chemicals will stick. When they stick, they cause the dendrite to open its sodium-potassium gates, which begins an electrical nerve impulse in that neuron. As soon as the chemicals have done their job, they are either dissolved or are recycled back into the terminal knob.

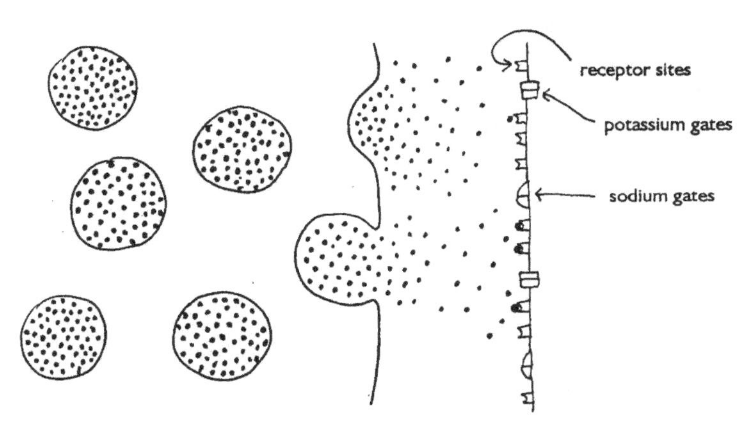


Understanding how synapses work allows you to understand how some drugs work. Poisons called ***neurotoxins*** can prevent the neurotransmitters from sticking to the sites they are supposed to stick to. For example, a poison called *curare* (kyur-ar-ay) made from the skin of a South American frog, does an even better job of sticking to these receptor sites than the neurotransmitter chemicals do. The curare gets in there and sticks to all the sites

so that there are none left for the neurotransmitters. This prevents the electrical signals from passing from neuron to neuron. No signals are sent to the heart telling it to beat. No signals are sent to the lungs telling them to breathe. You die quickly. There are also neurotoxins that do the opposite. They flood the gap so the signals won't stop. Opposite problem, but just as bad.

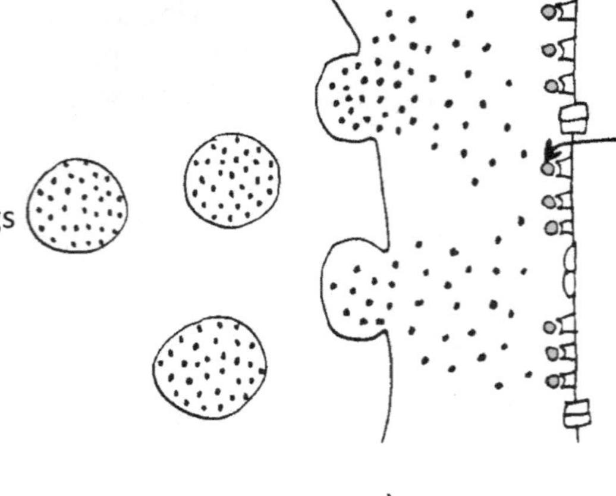

Uh-oh! There's already something on the site! (curare molecules) The neurotransmitters can't attach!

On the other hand, some chemicals can be a great help to the synapses. A medicine called diazepam can cause the chemicals to do a better job of sticking to the receptor sites, thus improving the functioning of the synapse. You could think of them as little dock workers, standing at the docking sites, helping the chemical ferry boats to get docked more quickly and safely. This medicine is used to treat anxiety and muscle spasms.

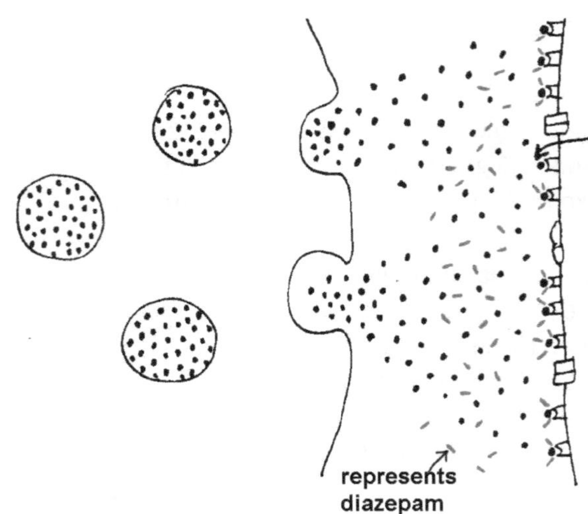

The neurotransmitters stick much better, thanks to the diazepam molecules.
represents diazepam

Another problem that can occur with the synapses is that the neurons do not seem to make enough neurotransmitter chemicals. Medicines called SRIs (serotonin re-uptake inhibitors) stop the body from dissolving the neurotransmitters so fast after they are used. (Serotonin is one of the neurotransmitters.) The SRI chemical slows down the rate of dissolving so that there are more neurotransmitters in the gap.

Why must the electrical signal be converted into a chemical signal, then back to an electrical signal between every cell? It seems that this allows the nervous system to be more responsive and more flexible. The nervous system is incredibly sophisticated and capable of fine-tuning itself to respond to the smallest changes. The synapse and the neurotransmitters somehow contribute to this sensitivity and adaptability. Some neurotransmitters even inhibit (try to stop) an electrical signal from getting passed along. This adds another layer of complexity.

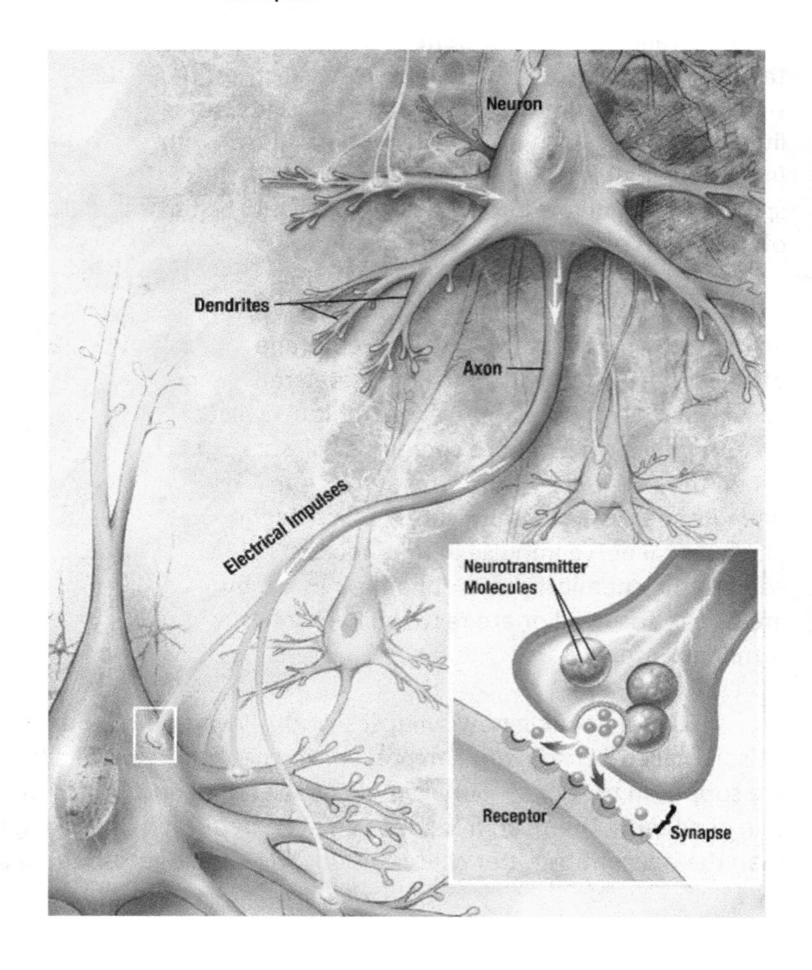

## ACTIVITY 5.5 REVIEW: What are these things?

See if you can remember what these things are. If you can't remember, try using the process of elimination. Do all the ones you are pretty sure of first, and save the harder ones for last. The possible answers are listed below.

1) ____________
2) ____________
3) ____________
4) ____________
5) ____________
6) ____________
7) ____________
8) ____________
9) ____________
10) ____________
11) ____________
12) ____________
13) ____________
14) ____________
15) ____________
16) ____________
17) ____________
18) ____________
19) ____________
20) ____________
21) ____________
22) ____________
23) ____________
24) ____________
25) ____________
30) ____________
31) ____________
32) ____________
33) ____________
34) ____________
35) ____________

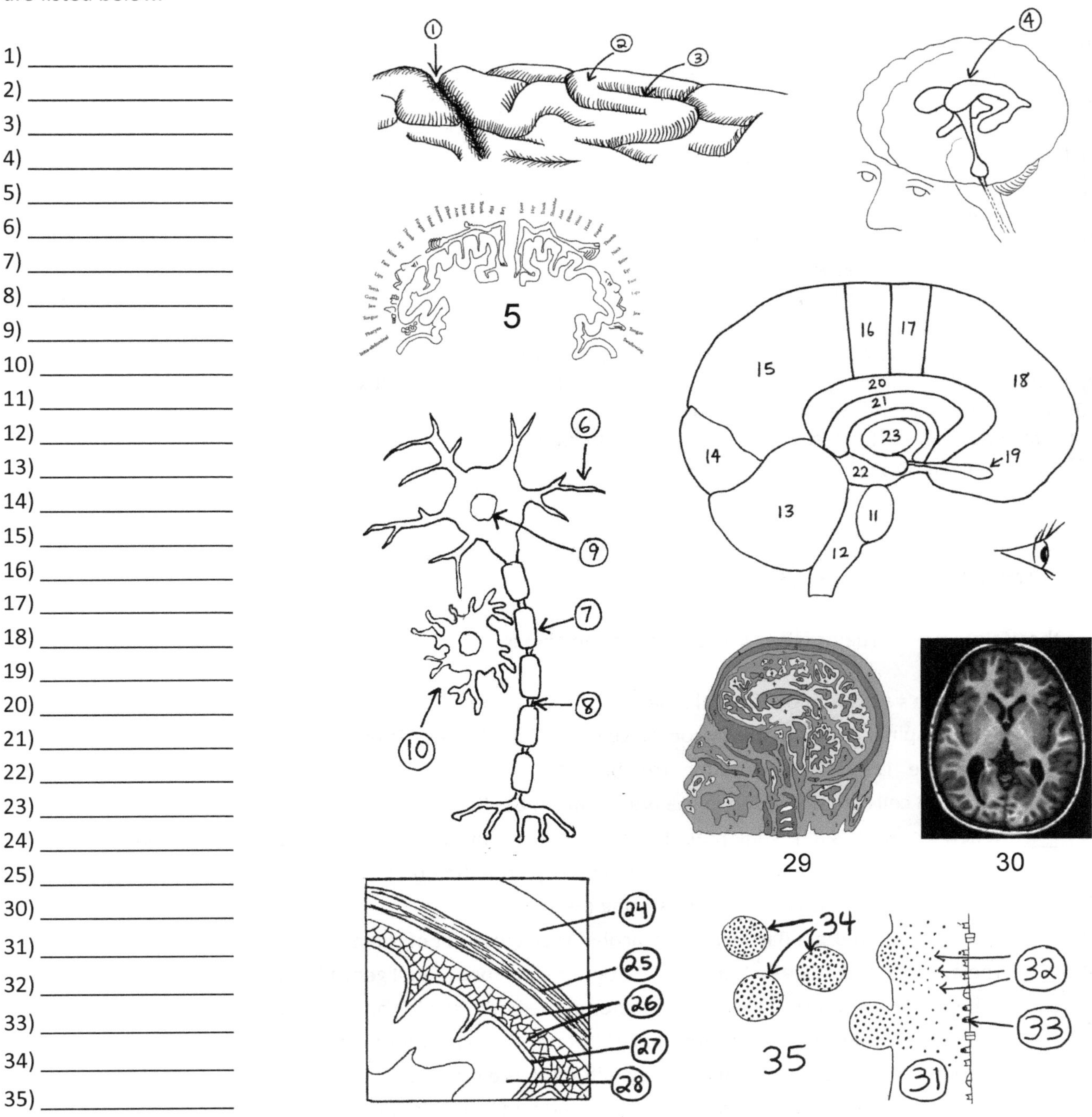


WORD BANK (possible answers): dendrites, pons, motor cortex, frontal lobe, gyrus, glial cell, ventricles, vesicles, node of Ranvier, arachnoid layer, dura mater, pia mater, cingulate gyrus, occipital lobe, nucleus, sulcus, dura mater, sensory cortex, corpus callosum, arachnoid layer, cerebellum, receptor site, olfactory bulb, Schwann cell, parietal lobe, thalamus, medulla oblongata, midbrain, fissure, homunculus, neurotransmitters, skull, cortex, PET scan, synapse, MRI scan, axon knob

## ACTIVITY 5.6 "Odd one out"

One of the words in each group doesn't belong with the others because it doesn't fit into the same category that the others so. Your job is to try to figure out what that category is (what most of the words have in common) and why that last word does not fit. For example, in this group of words: (heart, brain, bones, shoes) the word "shoes" doesn't fit because it is not a part of the body. In this group: (eyes, nose, stomach, ears) the word "stomach" doesn't fit for two reasons. It is not located on the head, and it is not part of your senses.

The first word groups are easier, and it gets a bit harder as you approach 10. Feel free to look back at the text to get clues!

1) limbic, temporal, frontal, occipital
2) fissure, gyrus, sulcus, synapse
3) PET, CAT, MRI, CRV
4) oligodendrocyte, astrocyte, microglia, neuron
5) oxygen, carbon dioxide, water, sugar
6) Schwann, Gage, Wernicke, Ranvier
7) fornix, hippocampus, mammillary bodies, medulla oblongata
8) frontal lobe, temporal lobe, olfactory bulb, occipital lobe, sensory cortex
9) temporal lobes, pituitary gland, ventricles, parietal lobes, olfactory bulbs
10) knob, synapse, dendrite, axon, nucleus
11) threshold, homeostasis, saltatory conduction, action potential
12) potassium. curare, serotonin, diazepam,

## ACTIVITY 5.7 TRUE or FALSE challenge

Are these statements true or false? Write T or F in the blanks.

1) _____ Wernicke's area is in the frontal lobe.
2) _____ Cerebrospinal fluid is made by special cells that line the ventricles.
3) _____ "Dendrite" is the Greek word for "tree branch."
4) _____ Schwann cells are not found in the brain, only in the body.
5) _____ When a neuron is at resting potential, it is more positive outside and more negative inside.
6) _____ A "shunt" is a tube that drains extra blood away from the brain.
7) _____ The hippocampus is necessary for storing memories.
8) _____ The BBB (Blood Brain Barrier) keeps alcohol molecules out of the brain.
9) _____ The amount of wrinkles a brain has is a better clue about intelligence than size is.
10) _____ The thalamus is sort of like a relay center in the brain, sorting and routing signals.
11) _____ Most capillaries in the body are "leaky" but not the capillaries in the brain.
12) _____ Cerebrospinal fluid carries neurotransmitters to the synapses.
13) _____ Astrocytes keep neurons from shifting around and losing contact with each other.
14) _____ There are two ventricles, one in each hemisphere.
15) _____ The hypothalamus is in charge of hunger, thirst, and body temperature.

# CHAPTER 6

## NETWORKS OF NEURONS

Neurons join together to form long, complicated networks. These networks could be compared to our systems of telephone wires, or electrical wires, criss-crossing the landscape, connecting all of our houses and businesses.

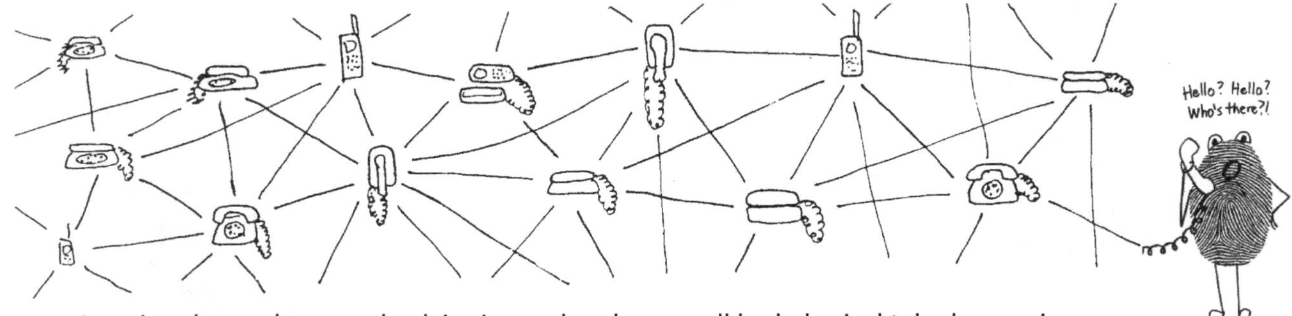


Imagine that we've gone back in time, when houses all had physical telephone wires going into them. (Many houses do still have a "land line.") Each telephone can represent a neuron. Let's have an exciting piece of news represent a nerve impulse. We'll let letters represent the telephone callers.

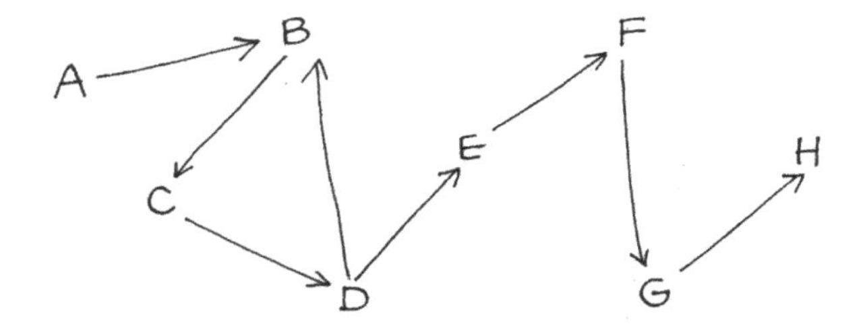


Person A calls person B, who calls person C, who calls person D, who calls person B (oops, sorry person B — you've already been called), who calls person E, who calls person F, and so on and so forth, until everyone knows the news.

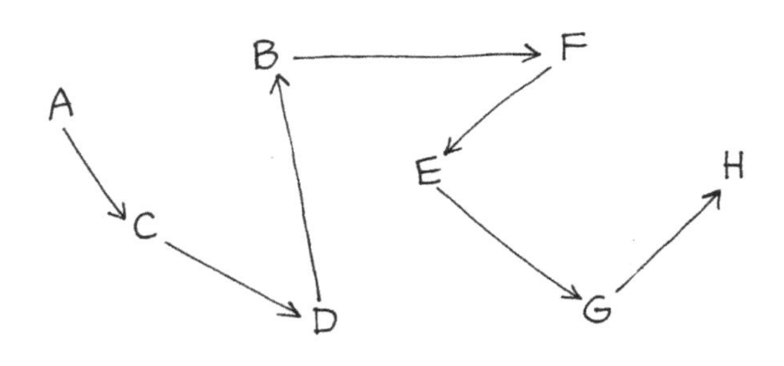


The news could have taken a different path, with the same result. Person A could have called person C, who called D, who called B, who called F, who called E, who called G, who called H.

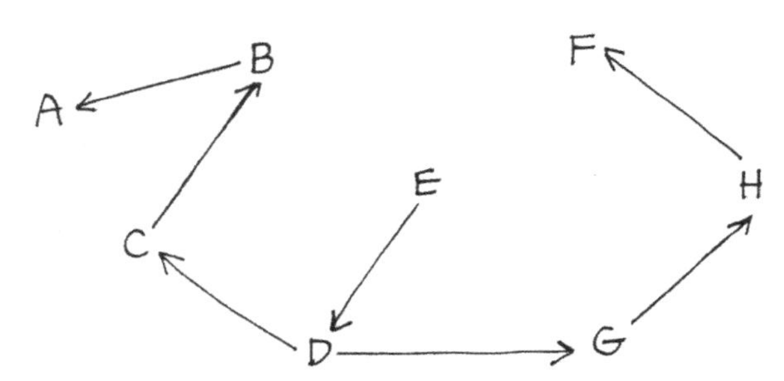


What if person E was the first to know? E calls D and person D makes two calls, to both C and G, and these two then call the rest.

As you can see, there are many possibilities for how the message could travel. The same is true in your nervous system, but on a grand scale: there are BILLIONS of different paths that a nerve impulse can take.

The path that a nerve impulse takes is called a ***neuronal network*** (or ***neural network,*** though the primary meaning of this term now seems to be digital networks using computers). There can be many networks within the same group of neurons. Each pathway is considered a separate network. A network involves thousands of neurons.

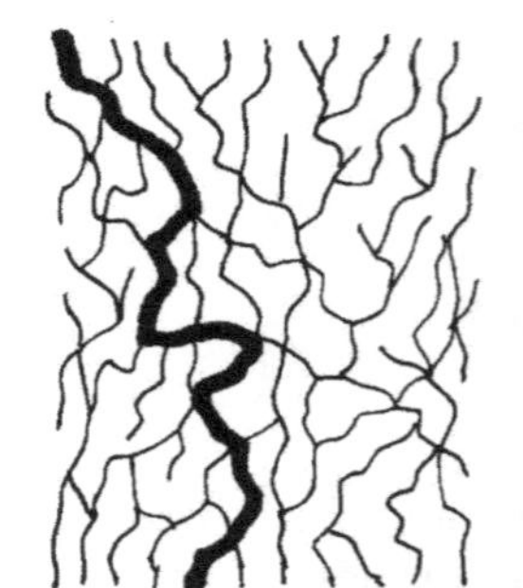 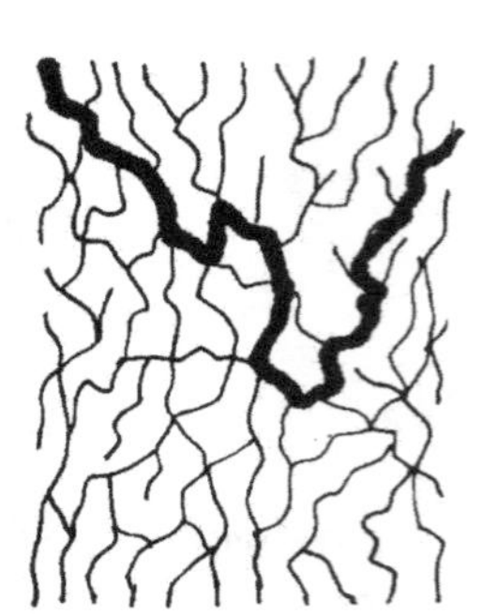 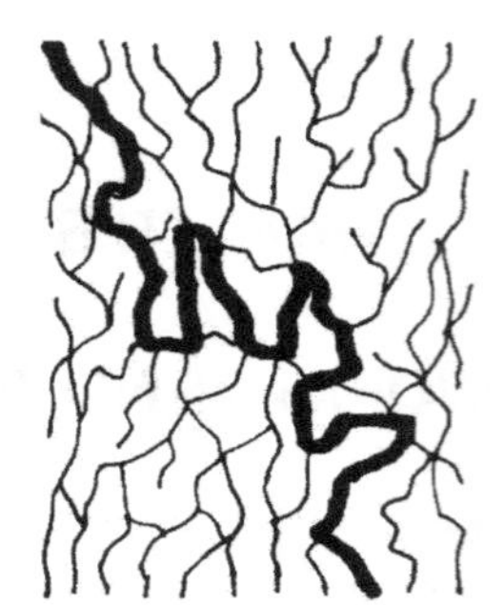 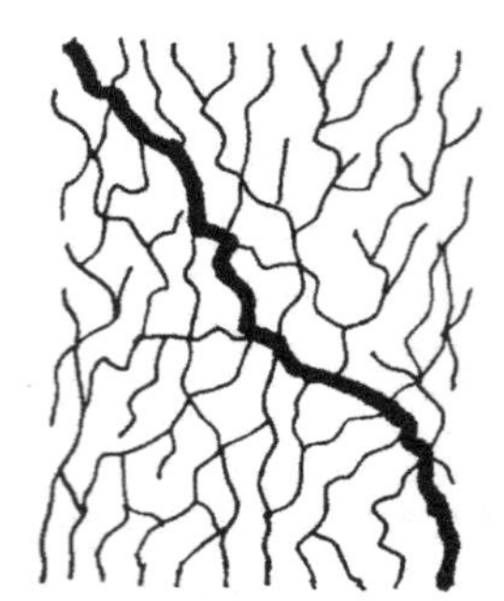 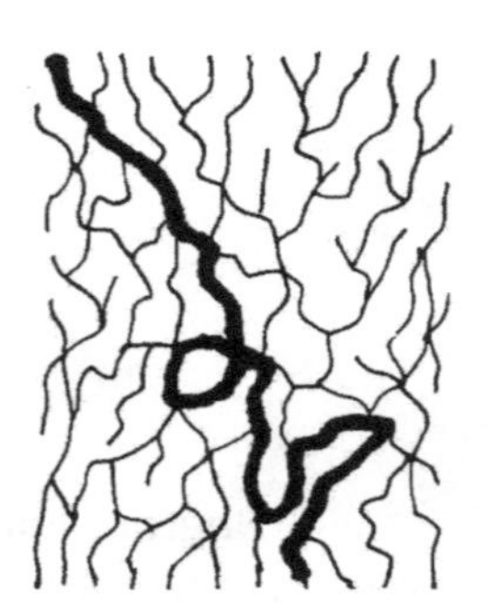

Each time your brain processes information of any kind, an electrical impulse traces out a pathway through the brain. The diagrams above show the same network of neurons (the lines would be dendrites and axons), but with five different pathways traced out. If exactly the same process is repeated, the nerve impulse will travel that route again. The more the pathway is traveled, the easier it becomes for the impulse to travel it. It's a little bit like walking through a field of tall grass. The first time, there is no path to follow and you have to work at getting through. You may have to beat down the grass with your feet, and walking seems difficult. The next time you walk through, it's a little bit easier because you can still see the bent grass where you went through before. By the tenth time you go over that path, the grass is permanently crushed down and a trail is starting to appear. By the 100th time you go across, the path has become easy to travel.

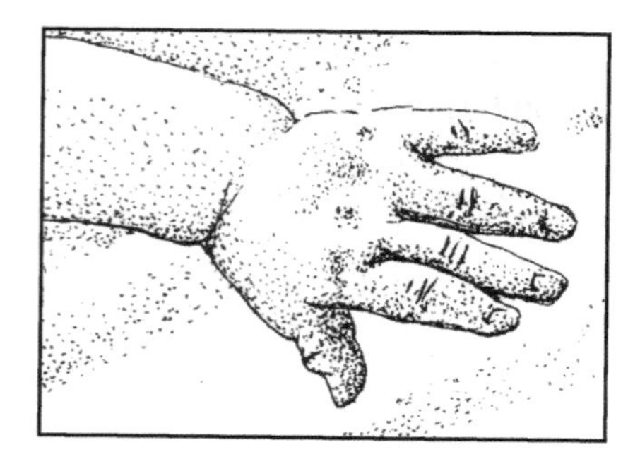

For example, the first time a baby tries to move his hands, the nerve impulse has no path to follow—it must find its own way across the brain. The second time the baby tries to move his hands, the nerve impulse can try to follow the pathway it first made. If successful, the nerve impulses will keep going the same route again and again, until the pathway is so easy to follow that the baby no longer has to concentrate on trying to move his hands—the motion becomes automatic.

The first time you try to play a song on the piano, it is very difficult. You have to think about every note and every movement of every finger. The nerve impulses are creating brand new pathways across the brain. As you play the song again and again, the nerve impulse retraces that path again and again. The pathway becomes more like a roadway. Soon, the impulses are zipping along effortlessly. You can play the song with very little effort.

Once a pattern of basic networks has been created, the brain can add more to them. For instance, once you know how to play simple songs on the piano, you are ready to start working on harder ones. Your brain will build on those first networks and add the new information about the harder songs. As the networks get more and more complex, you will be able to tackle more difficult songs. You can't play the difficult songs if you have not built the basic networks first, using the simple songs.

These network pathways can stay in place for a very long time, sometimes for your whole life. Many people find that skills they learned when they were young stay with them right through old age. Of course, if you only took piano lessons for a couple of months, don't expect your pathways to be as permanent as those of someone who took lessons for years. The pathways do need to be reinforced in order for them to become permanent. (In other words, practice makes perfect!)

Here is a diagram that shows a network becoming more complicated:

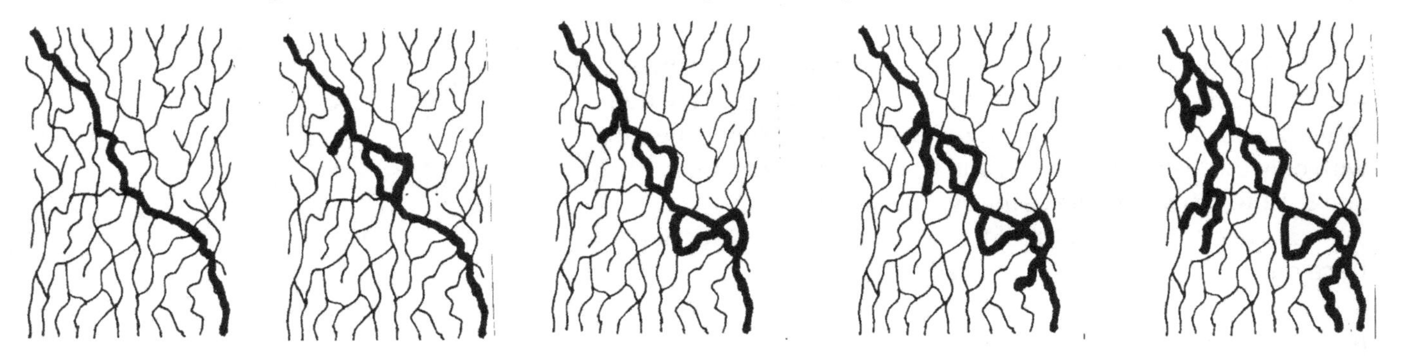

The networks we have drawn so far are just cartoons, not real ones. Real networks look very messy. You can't tell what's connected to what. This is more what a real network of neurons looks like:

The above image really is a photograph of neurons. The images below aren't photos of neurons, but the art is based on the idea of neuronal networks.

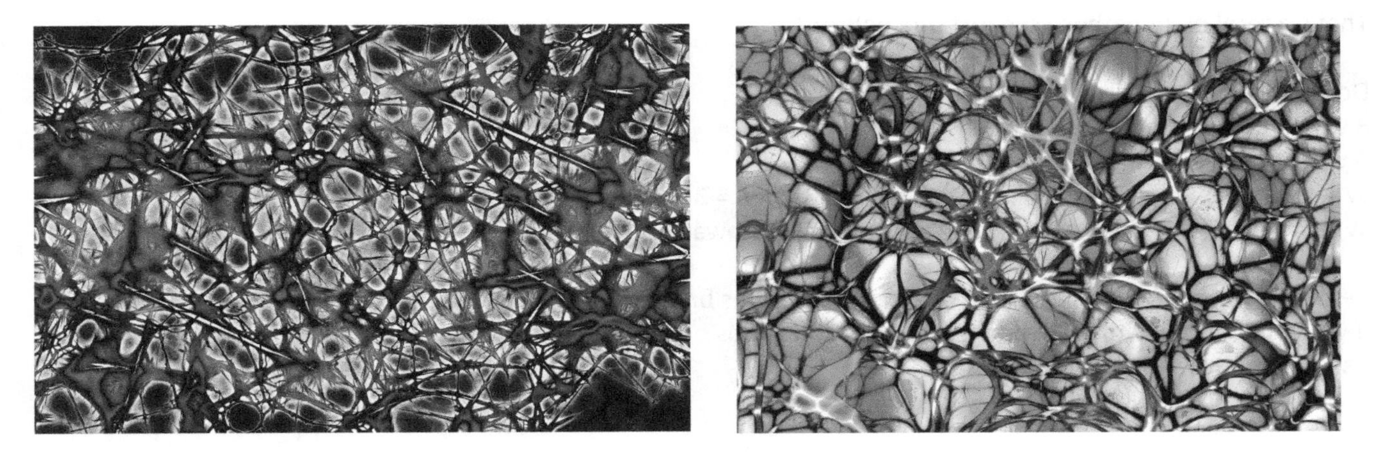

## ACTIVITY 6.1 Neuron network math

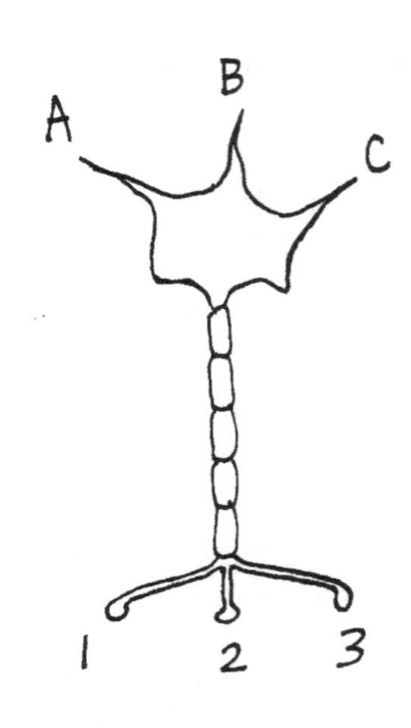


Even in just a single neuron there are different pathways that an impulse could take. It could go in and out any of the dendrites and axon knobs. Let's consider a small, very simple neuron with only three dendrites, A. B, and C, and three axon knobs, 1, 2, and 3. How many possible pathways would there be?

We could make a list of the possibilities. Use your finger to trace out each possibility as you read it.

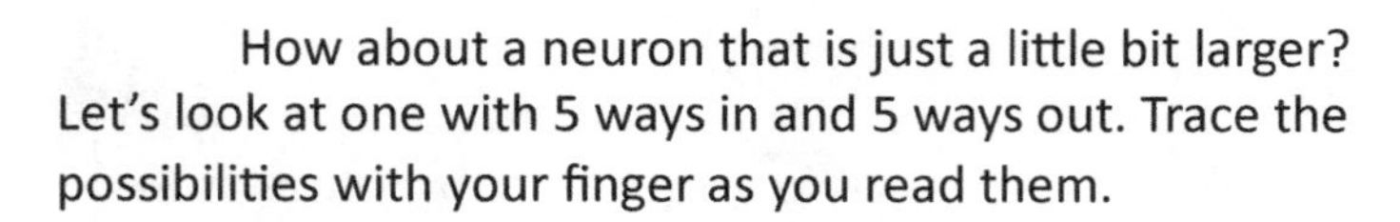

| | | |
|---|---|---|
| In at A, out at 1 | In at B, out at 1 | In at C, out at 1 |
| In at A, out at 2 | In at B, out at 2 | In at C, out at 2 |
| In at A, out at 3 | In at B, out at 3 | In at C, out at 3 |

That's a total of 9 pathways.

How about a neuron that is just a little bit larger? Let's look at one with 5 ways in and 5 ways out. Trace the possibilities with your finger as you read them.

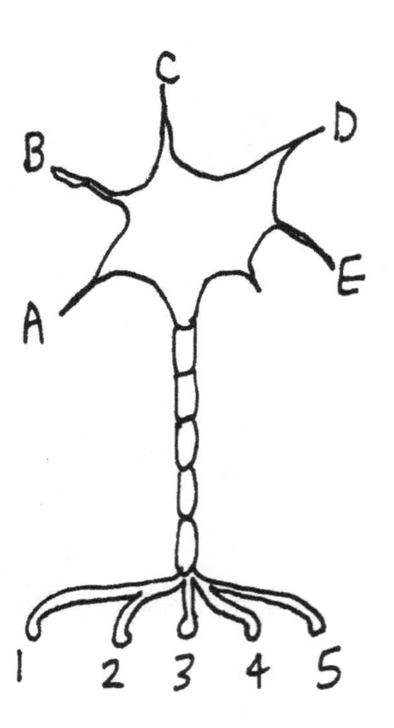


In at A, out at 1
In at A, out at 2
In at A, out at 3
In at A, out at 4
In at A, out at 5

In at B, out at 1
In at B, out at 2
In at B, out at 3
In at B, out at 4
In at B, out at 5

In at C, out at 1
In at C, out at 2
In at C, out at 3
In at C, out at 4
In at C, out at 5

In at D, out at 1
In at D, out at 2
In at D, out at 3
In at D, out at 4
In at D, out at 5

In at E, out at 1
In at E, out at 2
In at E, out at 3
In at E, out at 4
In at E, out at 5

That's a total of 25 pathways in just one neuron!

Do you notice a pattern?
3 ways in, 3 ways out, is 9 pathways;
5 ways in, 5 ways out, is 25 pathways.
What would 6 ways in and 6 ways out give you? 6 x 6 = 36 pathways.
What about 6 ways in and 7 ways out? 6 x 7 = 42 pathways.

Real neurons in the brain can have dozens of dendrites branching out. What if you found a cell with 25 dendrites and 10 axon knobs? That would be 250 pathways in just one neuron!

Now, what if we form a very small network made of just two simple neurons? Let's consider these two beautiful specimens. You are going to fill in the pathway possibilities this time. We'll get you started, though.

In at A, out at 1, then in at D, out at 4
In at A, out at 1, then in at D, out at 5
In at A, out at 1, then in at D, out at 6

In at A, out at 2, then in at E, out at 4
In at A, out at 2, then in at E, out at 5
In at A, out at 2, then in at E, out at 6

In at ___, out at ___, then in at ___, out at ___
In at ___, out at ___, then in at ___, out at ___
In at ___, out at ___, then in at ___, out at ___

In at ___, out at ___, then in at ___, out at ___
In at ___, out at ___, then in at ___, out at ___
In at ___, out at ___, then in at ___, out at ___

In at ___, out at ___, then in at ___, out at ___
In at ___, out at ___, then in at ___, out at ___
In at ___, out at ___, then in at ___, out at ___

In at ___, out at ___, then in at ___, out at ___
In at ___, out at ___, then in at ___, out at ___
I In at ___, out at ___, then in at ___, out at ___

In at ___, out at ___, then in at ___, out at ___
In at ___, out at ___, then in at ___, out at ___
In at ___, out at ___, then in at ___, out at ___

In at ___, out at ___, then in at ___, out at ___
In at ___, out at ___, then in at ___, out at ___
I In at ___, out at ___, then in at ___, out at ___

In at ___, out at ___, then in at ___, out at ___
In at ___, out at ___, then in at ___, out at ___
In at ___, out at ___, then in at ___, out at ___

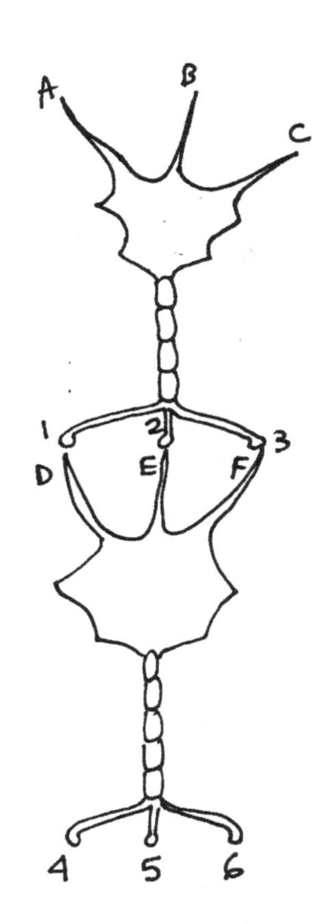

That's a total of 27 pathways, right? 3 x 3 = 9, and 9 x 3 = 27. [3 x 3 x 3 = 27]
The third (x 3) came from the additional ways out: 4, 5, and 6.

What if the axon knobs could go to any of the next dendrites? What if 1 could go to either D, E, or F, and 2 and 3 could each go to D, E, or F? How many possibilities would that make? Making a list for that one will take up too much space. Let's find a short cut. We know that one neuron with 3 ways in and 3 ways out will provide 9 pathways. So we have 9 pathways for the top neuron, and 9 for the bottom neuron. If we match up each of the top pathways with each of the bottom paths, we will get 9 x 9 = 81 pathways.

Would you like to calculate the possible pathways for the network shown here on the right? No? Okay, we'll let you off the hook as long as you ponder the following piece of information: Your brain contains about 100 billion neurons (100,000,000,000). The number of neuron networks possible in your brain is billions of billions—a number so high that you probably don't even know a name for it. Just think of how much potential you have—make the most of it!

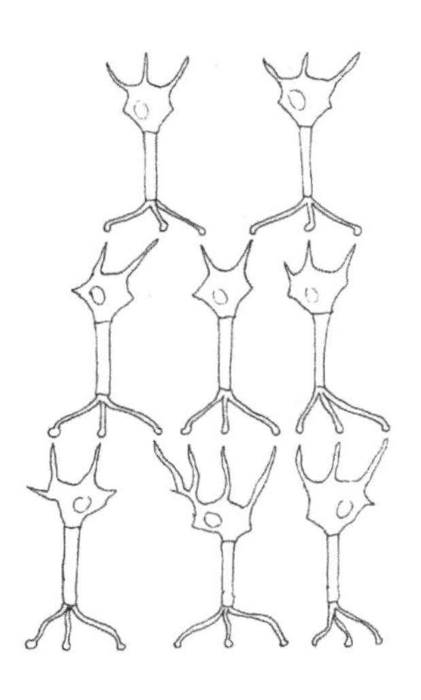

nglia
16)

## ACTIVITY 6.2 Reinforce some of your neuronal networks

We learned in this chapter that the more you reinforce the neuronal networks in your brain, the longer they will last. You've been creating new networks in your brain as you have been learning about the brain. Let's go back over some of the information you have learned and see if those networks are still in place.

Can you remember the six basic lobes of the cerebrum?

1) ____________________ 2) ____________________

3) ____________________ 4) ____________________

5) ____________________ 6) ____________________

Label the drawing with the numbers of the lobes.
Label the brain stem as number 7 and the cerebellum as number 8.

7) What is the brain stem in charge of? ____________________

8) What does the cerebellum do? ____________________

9) The hippocampus was named after this animal: ____________

10) What brain part acts like brakes, inhibiting (slowing down or stopping) motions? ____________

Use your memory of "The Brain Song" to fill in these blanks:

11) *"Without my ________ to get me up, I'd sleep the whole day through."*
12) *"Oh, what a sight for my ____________________."*
13) *"Pans and dishes filled the sink. My ____________ lobe could smell the garbage stink."*
14) *"I fled that great disaster using __________ __________ nerves."*
15) *"My ____________ lobe decided to take me out the door."*
16) *"My ____________ lobe could hear the birdies sing."*

17) The part of your brain that makes you feel hungry and thirsty is the ____________________.
18) The olfactory bulb connects the brain to the ________.
19) One of the pituitary's jobs is to make you g______.
20) The neuron's "branches" are called the ________________.

If the statement applies to the left hemisphere, put an "L" in the blank.

If the statement applies to the right hemisphere, put an "R" in the blank.

21) Very good at logic and mathematical reasoning: ____
22) Creative and musical: ____
23) Sees the "whole picture" better than the details: ____
24) Good at drawing and sculpting: ____
25) Uses words to name things: ____

# CHAPTER 6.5

## MORE ABOUT NEURONAL NETWORKS

Nerve impulses trace out particular pathways, making unique networks in the brain. These networks store information as simple as knowing that sandpaper feels rough, to more complex facts such as 8x7= 56. Networks are formed when our bodies learn to do certain tasks, such as riding a bicycle or typing on a keyboard. Even our limbic system forms networks of emotional patterns. We have networks for things like social customs and what kind of clothes we find fashionable. Neuronal networks are the way our brains work.

What would happen if a brain formed networks based on faulty information? What if you learned your multiplication tables all wrong? How long would it take for your brain to cancel those old networks and build new ones for the right answers? Or what if you grew up in a neighborhood full of mean dogs. How many good experiences with dogs would it take to convince you that most dogs are friendly? Erasing old networks is often more difficult than building new ones. Networks die hard, and re-routing nerve impulses is hard work.

Have you ever moved to a new house? If so, you know that for the first few weeks or months in your new house, you have to think about where to find things. If you want to find your tennis racket, for example, the first thing that pops into your mind is where the tennis racket was stored in your old house. Then your frontal lobe has to say, "No, that's not right any more. We need to form a new network for the location of the tennis racket." Then you think about where you put the racket when you moved into your new house. The act of thinking about it forms a new network. If you chose a storage place in the new house that is very similar to the storage place in the old house, such hanging on the wall in the garage, this can make it easier to form a new network, because you can use part of the old network. Your old network knew that tennis rackets hang on walls in garages. If this is still true, then you can use that part of the old network and just replace your mental image of the old garage with the new one.

I can't find anything in this new house!!

What happens when huge neuronal networks are not needed any more? Do they disappear right away? No, and in fact, they probably don't ever get erased entirely. If you move to a different country and learn a new language, you won't forget your native language. Those networks will stay there unless you were very young and the networks were not firmly in place yet. If you stop practicing the piano, over the years your brain will lose some of those connections, but you will probably never forget basic scales and simple songs. On a sad note, if you lose a friend or family member, your neuronal networks that allowed you to have a relationship with that person don't go away. The networks are still there, and not being able to use them any more will make your limbic system generate feelings of sadness and loss.

The phase of life when the greatest number of networks are formed is childhood. We make networks during our whole life, but not to the extent that we do when we are young. Children's brains develop networks more quickly, making as many as 1000 connections per minute. For example, the best time to learn a second language is between age 2 and age 12. It's not impossible after that, but it is much more difficult. (Strangely enough, most American schools don't offer foreign language to students under the age of 12. Hopefully, that will change in the future.)

Now, don't confuse making new networks with making new neurons. In general, you don't grow new neurons. (Recent research has found that neurons can regenerate more than we thought they did, especially in a few particular areas of the brain, but the general principle still holds.) Even children don't grow new neurons. By the time a baby is born, it has pretty much all the neurons it will ever have (over 100 billion of them!). It's the connections between those neurons that increase and grow throughout life.

A well-known brain researcher named Marian Diamond (at UC Berkeley) did experiments with rat brains to show how networks develop and grow. She wanted to see how much effect an animal's (or human's) environment has on the structure of its networks. Dr. Diamond designed an experiment where she raised groups of rats in different environments, then looked at their brains to see what happened to the neurons. She had four groups of rats. The rats in the first group were all by themselves and had no toys. The second group were all by themselves, but had toys. The third group had other rats in with them, but no toys. The fourth group had both toys and other rats with them. After the rats died, she opened up their brains and looked at the neurons under the microscope. What she found looked something like this:

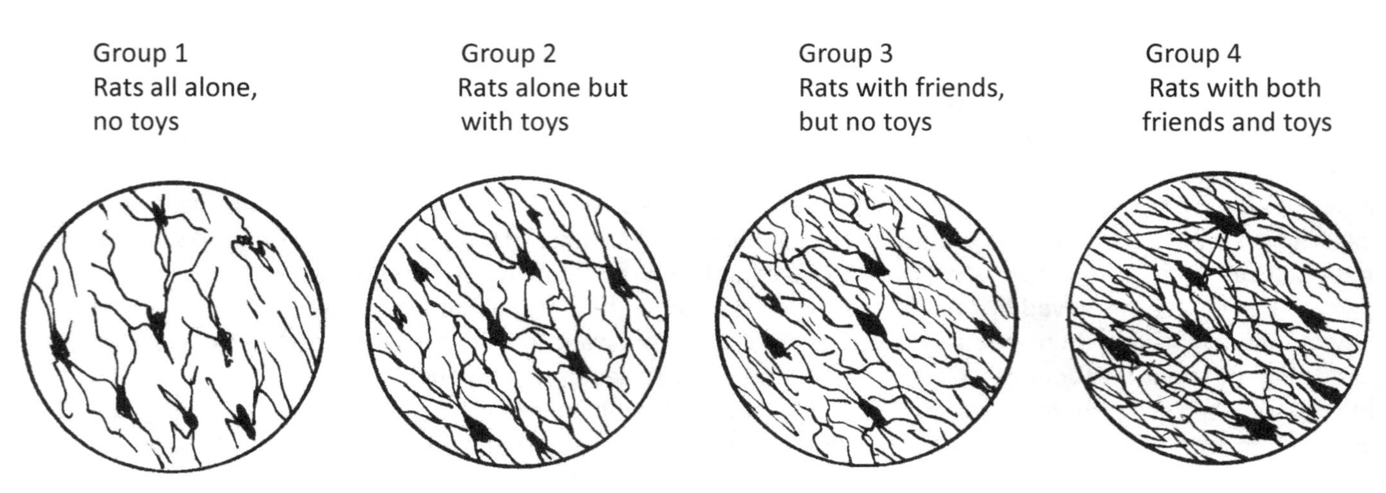


NOTE: These are not Dr. Diamond's actual slides. They are facsimiles (fak-SIM-al-ees) designed to illustrate the general idea of what she found. She counted the number of dendrites in certain areas.

Dr. Diamond concluded that environments affect the complexity of neuronal networks. She found that the rats in group 4 were just plain smarter. They learned mazes more easily and could adjust to changes in their environment with less stress. Other researches who have done similar experiments have found that the number of glial cells increases, as does the number and the width of capillaries (those tiny blood vessels that bring the oxygen and food to the cells). The number of energy-producing mitochondria inside the neurons can also increase.

Another special feature of neural networks is something called ***plasticity***. You can see the word "plastic" in there. Plastic is a substance that will do a lot of bending before it will break. Some plastics are pretty soft and can be reshaped if they are heated a bit. Plasticity in the brain means the ability to change and adapt when necessary. Networks can often be rebuilt if they are damaged. In fact, the brain is really good at re-routing nerve signal "traffic" around damaged areas. New networks will form to replace old ones. For example, when someone has a stroke, part of their brain does not receive enough oxygen and some of the cells die. Those cells were in networks. What happens then? The signals can't travel their usual paths. The brain sends out signals to move a hand or to say a word, and nothing happens. Fortunately, over time, and with a lot of therapy, new networks can form (Perhaps, we learned in the previous chapter, some ependymal cells will migrate to the damaged area and turn into glial cells or neurons.). Stroke patients often recover to an amazing degree, and their lost abilities will gradually return. Even older brains still have an amazing amount of plasticity.

Speaking of plastic and brains, I've got a plastic brain.
Bet you just bought yours online. I made my brains - one is Jello and the other is crocheted yarn!

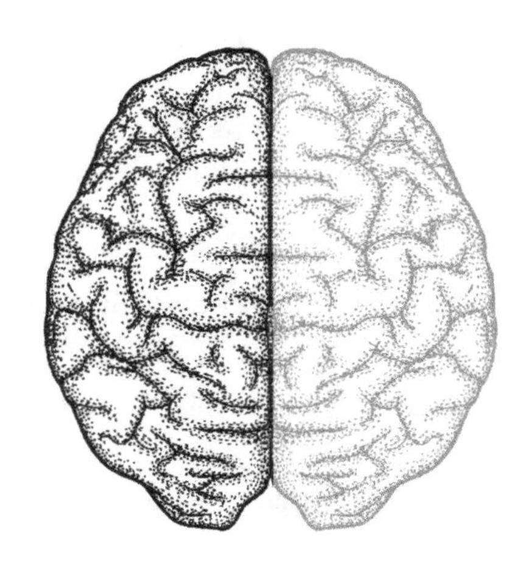

The most dramatic example of plasticity is when a child has to have an entire hemisphere surgically removed in order to stop the progression of a disease. This type of operation is called a ***hemispherectomy***. ("Ectomy" means "removing a part.") Can you live with only half a brain? The first surgeon to attempt this daring procedure was probably pretty nervous! However, his patient would certainly have died without the operation, so he went ahead. Amazingly, the answer is yes. Children can adapt and survive quite well with only one hemisphere. The remaining hemisphere gets busy and starts growing billions of new connections. It isn't long before the child is walking and talking again. This works for children, but what about adults? Adults have a lot of plasticity in their brains, but not as much as children do. Adults would not make such a good recovery after this type of operation. Probably they would have permanent paralysis on one side of their body (the side that had been controlled by the hemisphere that was removed). For this reason, surgeons won't perform this operation on adults.

What if the reverse happens—if a body part is removed but the brain parts it was wired to remain intact? Those brain parts are now "out of a job," so to speak. What do they do now? The same as anyone else—find new jobs! The neurons will try to rewire to other nearby networks. In the sensory cortex this can cause interesting results. A soldier who had lost an arm reported to his doctors that every time he shaved his face he had the sensation that his missing arm was tingling. The face and arm areas in the sensory cortex are close together, so it is likely that the arm neurons were sneaking over to the face area and "borrowing" incoming sense signals.

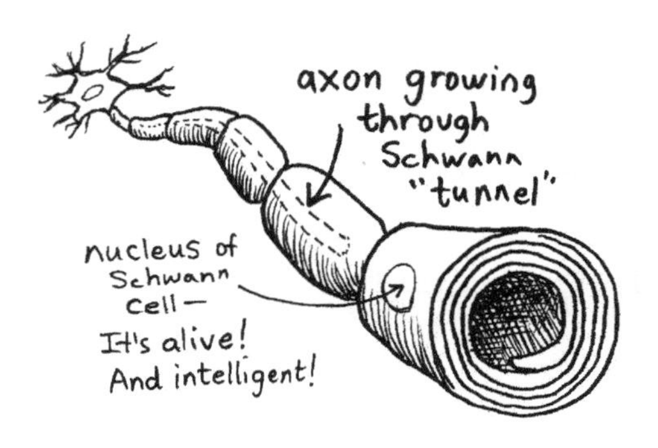


How does rewiring actually occur? This is a hot topic in brain research right now. One big discovery was that those Schwann cells aren't dummies. If an axon is severed, the Schwann cells form a tube through which the axon can regenerate itself. The Schwann cells somehow know in which direction the tubes should go, so the neurons can reconnect with other neurons. This process is a bit slow, though. Nerve damage can take months, or even years, to heal. But your body will try. Unfortunately, the body is often not able to overcome major trauma to the nerves in the spinal cord. Researchers are working hard to find a way to help spinal cord nerves reconnect.

* * * * * * * * * * * * * * * * * * * * * * * * * * * * * * * * * * * * * * * * * * * *

### ACTIVITY 6.3 Meet some kids who had a hemispherectomy

You can "meet" some kids who are living with only one hemisphere (on the YouTube playlist). Most of them had to have this surgery because of a type of epilepsy that caused severe, life-threatening seizures. (A seizure is like an electrical storm in the brain.) After their operations, most of these kids are completely free of seizures. Some still have trouble with seizures, but they are not as serious and occur less often.

If you want to "meet" a surgeon who does hemispherectomies, there is an interview and a full length movie (listed at the end) about Dr. Ben Carson, who is probably the most well-known brain surgeon alive today. Not only has he done amazing surgeries (like separating conjoined twins) but his personal story is also amazing. He rose from very difficult circumstances and through hard work combined with faith, managed to achieve great things.

While you are on the playlist, also check out a very short video about brain research on rats titled, "What happens when you tickle a rat."

## ACTIVITY 6.4 Decipher the punch lines

Use the clues to write the words in the boxes. Then use the numbers under the boxes to fill in the letters that will give you the punch line to the jokes.

The space between two Schwann cells is a: ☐☐☐☐ ☐☐ ☐☐☐☐☐☐☐
19 16 6 10

In the brain, these cells make myelin sheaths: ☐☐☐☐☐☐☐☐☐☐☐☐☐☐☐☐
13 12 3

These cells make tunnels for injured axons: ☐☐☐☐☐☐☐
1 22

Chemicals that cross synapses are: ☐☐☐☐☐☐☐☐☐☐☐☐☐☐☐☐☐
9 20 5 17 18

The ends of the axon look like: ☐☐☐☐☐
4

The insulating substance along axons: ☐☐☐☐☐☐
7 2 23 21

The places where neurotransmitters stick: (on the post-synaptic membranes, pg 26.4) ☐☐☐☐☐☐☐☐ ☐☐☐☐☐
15 14 8 11

*What did the Hollywood movie producer say to the little neuron who auditioned for a part in his movie?*

☐☐☐, ☐☐☐, ☐☐☐'☐☐ ☐☐☐ ☐☐☐☐☐☐☐☐☐
1 2 3, 4 5 6, 7 8 9 ' 10 11 12 13 14 15 16 17 18 19 20 21 22 23

The ability of the brain to rewire after injury: ☐☐☐☐☐☐☐☐☐☐
9 8 1 11

The brain has over 100 _________ neurons: ☐☐☐☐☐☐☐
4

A type of neurotransmitter (starts with S): ☐☐☐☐☐☐☐☐☐
3 7 13 10

A group of neurons that function together: ☐☐☐☐☐☐☐
5 12 14 15

The brain structure that stores memories: ☐☐☐☐☐☐☐☐☐☐☐
2 6

*What is the brain's favorite television channel?*

☐☐☐ ☐☐☐☐☐☐ ☐☐☐☐☐☐☐
1 2 3 4 5 6 7 8 9 10 3 11 12 13 14 15

# CHAPTER 7

## LEARNING AND MEMORY

When our brain forms a new network between neurons, we call this ***learning***. You probably think of learning as just memorizing spelling words or math facts. A neuroscientist would say that every experience your brain records is learning, even if it is something simple like the feel of snowflakes falling on your hand, or the sound of someone's voice. Our brains are constantly learning, and most of this learning does not take place in a classroom. Every little piece of information that our brain takes in creates a unique pathway through the brain.

Now we are ready for the next step: networks of networks! All the individual networks are parts of larger networks that we call ***memories***. A memory is a group of networks that work together. Each one of the individual networks can be in a different part of the brain, yet they function together to form a single memory.

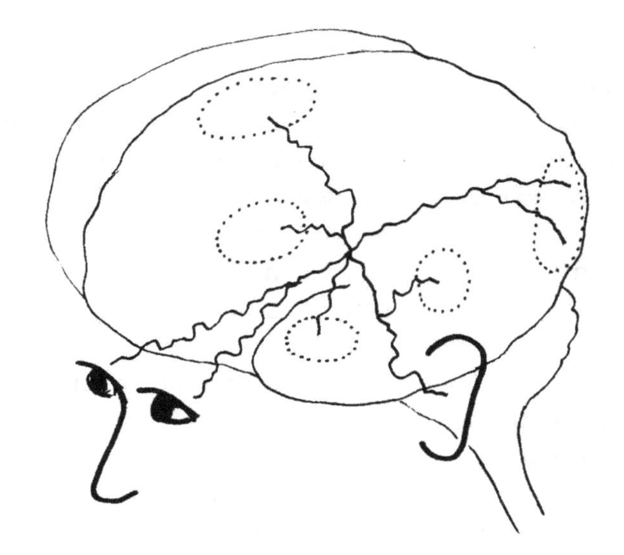

Let's use a specific memory as an example: your most recent visit to your favorite restaurant. The network for the shapes and colors of the walls and furniture went from your eyes to the outer part of your occipital lobe. The delicious smells went from your nose into your temporal lobe. The voice of your friends and family went from your ears through a network that ended up in the cortex of one or both temporal lobes (but in a different area from the smells). The meanings of the words your heard were stored in yet another area of the cortex, not the same one that the sounds went to. Your feelings as you entered the restaurant formed a network in the area that stores emotions. In order to remember this experience, all of those individual networks have to work together to bring back the complete memory. The visual network, the smell network, the sound network, the meaning network, and the emotional network all have to fire at the same time to recreate that memory.

This is very similar to how a computer works. It stores information in various parts of the disk or drive. To bring a certain text back onto the screen, it must find all those pieces of information and bring them back together again. When the computer starts running out of large storage space, it must start stuffing tiny bits of information in any "crack" it can find. This slows it down. We say the information is becoming "fragmented" and it's time to "defrag" your computer.

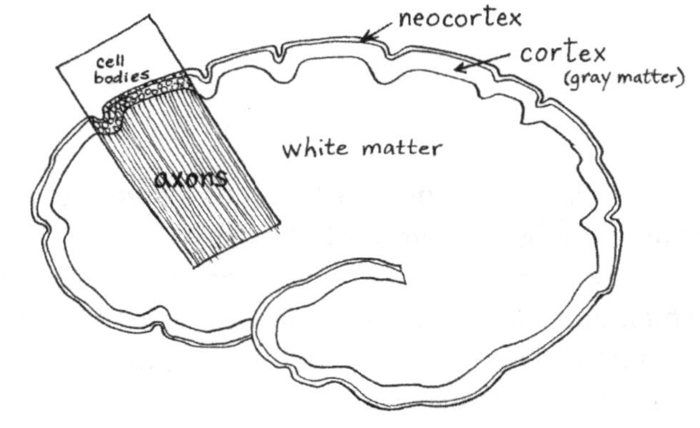


The part of your brain that stores memories is the very outer layer of the cerebrum, called the cortex, or the "***gray matter.***" (You remember that from chapter 2, right?) The very outer layer of the cortex is called the ***neocortex***. Your neocortex is like the available disk space on your computer. However, unlike your computer, you will never run out of space. Brains have an amazing capacity to store information.

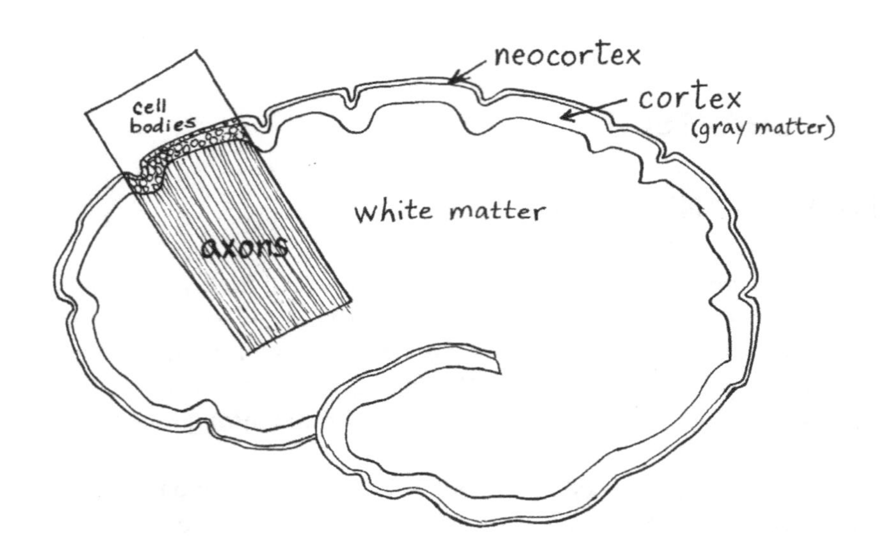


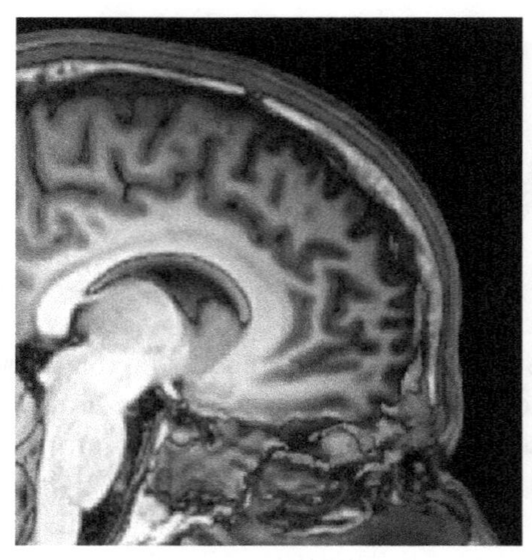
An MRI showing gray and white matter.

The gray matter is gray because it is made of the cell bodies of neurons. The cell bodies are darker in color than the axons. In a living brain, the gray matter is more of a pinkish-tan color because of all the blood. The gray matter doesn't appear truly gray until the brain is disconnected from its blood supply and preserved in chemicals. The white matter is white because it is made of axons that are covered with myelin, which is very light in color. There is no myelin in the gray matter.

Neuroscientists are not sure exactly what happens in the neurons to make a memory permanent. Some say it is a chemical change in the body of the cell. Others say it is a change in the dendrites. One of the most recent theories suggests that there are important changes to the internal protein"skeleton" that allows cells to maintain their shape. To make it even more mysterious, there are people who have died on the operating table, and then been revived, who have come back with memories of hearing and seeing things taking place in the room while their brains were registering zero brain activity! If someone tells you that scientists have figured out exactly how memory works, don't believe them. There are many things about memory that we don't yet understand.

However, even though we don't completely understand how memory works, we do know some basics. Let's take a look at these basics.

`We can classify memories into these categories: ***sensory, short-term, and long-term.***

Every memory starts out as a ***sensory memory,*** and only last a few seconds. Your senses are constantly being bombarded by sight, sound, smell and touch sensations. Your brain can keep track of up to 12 of these for just a few seconds. It then sends them through a screening system that determines which ones are important and which are not. This screening system isn't something you think about, it's automatic. Only the sensory information that is really important, or that "catches your attention," is sent on to the short-term memory.

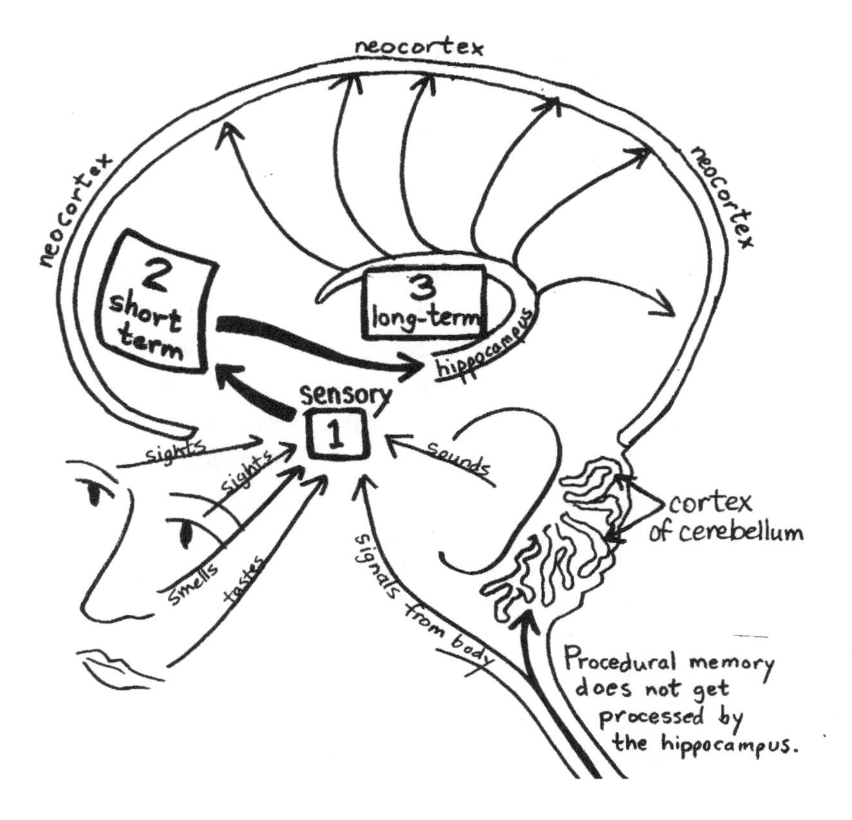


Your ***short-term memory*** lasts a bit longer than the sensory, but it still lasts for only a minute, at most. The short-term memory is sort of like a scratch pad or clip board, where temporary notes are posted. This scratch pad, located in the very front of your frontal lobe, can only hold about seven items (absolute maximum of nine), no matter how intelligent you are or how good you think you are at remembering things. (Even Einstein's short-term memory was the same as yours.) When the capacity of seven items is reached, your brain must erase the first ones to add more. If you want to keep any of the items longer than one minute, your brain must transfer them to ***long-term memory***.

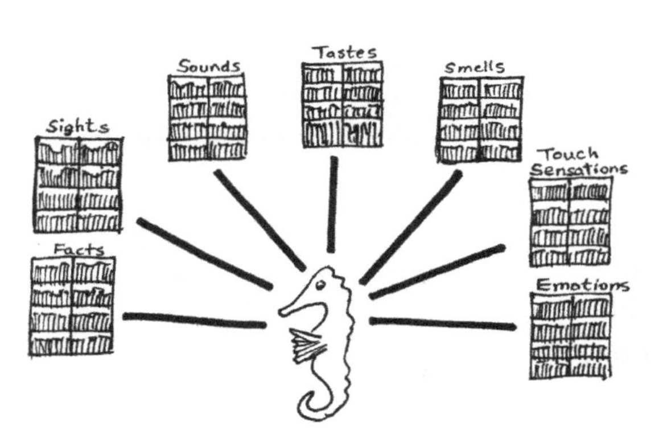


One of the brain parts responsible for transferring short-term memory to long-term memory is the ***hippocampus***. The "sea horse" librarian is responsible for taking short-term memories from the prefrontal cortex behind your forehead and transferring them to your permanent memory storage in the outer layer of your cortex, the ***neocortex***. The hippocampus acts as a librarian sorting out incoming books. It classifies each piece of short-term information as sight, sound, smell, taste, touch, meaning, or emotion, and sends the pieces on to the appropriate storage area in the neocortex. It can reverse this process, too, and bring the bits back together to recreate the memory. Then, when you have finished remembering, it puts everything back.

You may have noticed that the hippocampus is right near the ***amygdala***. The amygdala is in charge of strong emotions. This placement is not accidental. The close association between these brain parts causes you to remember things better if there is strong emotion attached to them. This is why you remember so clearly conversations you had when you were angry, or what the weather was like on a day that something terrible happened to you, or what you were wearing when you embarrassed yourself in front of your friends. Have you ever found that answers to questions that you get wrong, stick in your head better than the ones you get right? Strange as it might seem, that fear of failure actually helps you remember better. And that embarrassing moment with your friends—your hippocampus will make sure that you never forget not to do that again!

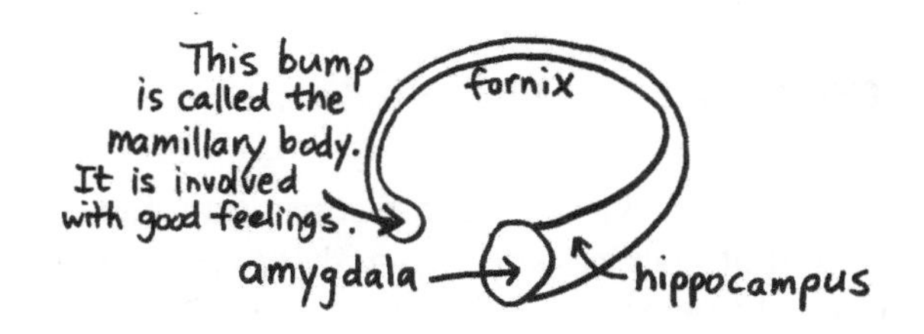

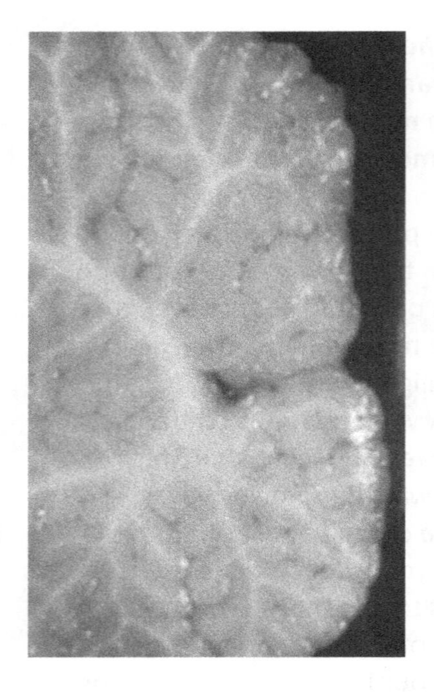

One special part of long-term memory is stored not in the cerebrum but in the cerebellum. The cerebellum also has an outer cortex that can store information. Since the cerebellum is in charge of coordinating body movement, it is in the cerebellum that you remember how to ride a bike, do a dance, or play notes on a musical instrument.

We know this partly because of a modern-day Phineas Gage—a man whose extremely improbable injury has let brain researchers discover some of the secrets of how the brain works. Clive Wearing lives in England. He used to be a well-known musician and symphony conductor. One day, in 1985, he came down with a virus. Usually this virus only causes cold symptoms or small sores in and around the mouth. For some unknown reason, the virus traveled up to his brain (got past the blood-brain-barrier) and began attacking his hippocampus. In a matter of days, he began losing his ability to transfer short-term memories to long-term storage. Within a few weeks he was reduced to having a seven-second memory. After a few years, he improved slightly and now he has about a 30-second memory.

Can you imagine what that would be like? Every thirty seconds your brain would totally "reset" and you would have no memory of anything that happened even one minute ago. If someone left the room and came back a minute later, it would be as though they were just coming in for the first time. If you tried to watch a television program, you wouldn't be able to remember anything that happened before the commercial break. If you tried to cook, you'd forget the recipe halfway through. Life would be impossible!

Brain researchers have been studying Clive, and have done MRIs to see what parts of his brain were damaged. The most amazing thing they have discovered is that he has retained his memory of how to play music. He can sit at the piano and play very difficult pieces that he had memorized before his accident. He can also play cards, tie his shoes and ride a bicycle. This tells us that the parts of memory involved in bodily actions do not need to go through the hippocampus. Clive, with no hippocampus, can still play the piano just like he always did. The cerebellum can remember **procedures**, such as how to play Solitaire, or how to put your fingers on the piano keys to play a song. This type of memory is called ***procedural*** memory.

Now I'm scared! I'm afraid I'll catch a virus and it will go to my brain!
Don't worry! Clive's case is extremely unusual. The chances of it happening to you are almost zero!
You really don't have to worry about it!



To conclude our story of Clive, don't feel too bad for him. He lives in a very nice house specially designed to take care of brain injury patients. The staff is trained to deal with his special needs and they take good care of him. His wife comes to visit him and he always greets her with hugs and kisses and says how much he loves her. She still loves him, too. (You can "meet" Clive through a video posted on The Brain's youtube channel.)

### ACTIVITY 7.1 Meet people with world-class memories

The playlist has some short interviews with world-class memory champions. They all use the Roman "loci" method of memorization (in activity 7.3). The first video shows a man named Andi Bell memorizing (in less than an hour) the location of every card in 10 decks. All of these champions say they started out with normal memories or even poor memories. They'll all tell you they are ordinary people who just decided to improve their memories.

**ACTIVITY 7.2 Watch a slide show on "The Anatomy of Memory"**

This online slide show is presented by the Exploratorium, a hands-on science museum in San Francisco, CA. This presentation shows dissection of a sheep brain. It is not gross at all. They just show you clean-looking slices of white and gray, and a picture of the inside, showing the corpus callosum, brain stem, thalamus, etc. No blood, no gore, just neat and tidy white and gray parts. And remember, it is a slide show with only still pictures. If you want a video version, they have one, but the picture quality is not nearly as good as the slide show.

**http://www.exploratorium.edu/memory/braindissection/index.html**

**ACTIVITY 7.3 Mnemonics** *(Nem-ON-iks)*

Tricks that help your brain convert short-term memory to long-term are called mnemonics (the first "m" is silent, so you say "nemonics"). A mnemonic helps the brain by associating a new piece of information with something already in long-term storage.

The ancient Romans were the first people known to have discovered mnemonics and make use of them. Roman politicians or philosophers would often give very long lectures or speeches and apparently worried that they would forget what they wanted to say. So they devised a system they called "loci," meaning location. The Roman orator would imagine a place he knew well and see each thing he was to remember as a part of that location. For example, if the first three locations he saw in his mind were a couch, a table, and a doorway, and the first three objects he had to remember were a horse, a sword, and a soldier, he might imagine the horse lying on the couch, the sword thrust through the center of the table, and the soldier standing in the doorway.

Mnemonics can also be used to help us spell difficult words. Making up a rhyme or a funny comment about a word that using some of the letters in the word can help us to remember the all the letters. For example, "There's A RAT in sepARATe" can help you to remember that it's spelled with an "a" in the middle, not an "e." Some people use a little sing-song rhyme to help them remember how to spell the state of Mississippi: "M-ISS-ISS-IPP-I" If you have trouble remembering how to spell the word "friends" you might make up something like this: "With school friends, FRIday ENDS the week."

Now it's your turn! Choose three words that you have trouble remembering how to spell. (If you don't have any, then think of a word that a friend or sibling has trouble with.) Make a mnemonic for each of them. Write the spelling word on the line and the mnemonic in the space to the right. (The mnemonics can be silly as long as they work!)

1) ________________ Mnemonic:

2) ________________ Mnemonic:

3) ________________ Mnemonic:

## ACTIVITY 7.4 Write about your earliest memory

Can you remember anything from when you were four years old? How about three? If you can remember something from when you were two, you are truly remarkable. If you say you can remember something from when you were one, we won't believe you because at that age your hippocampus was not developed enough to put things in long-term storage. Babies cannot form long-term memories. If someone ever tells you they remember being born, don't believe them. It's scientifically impossible!

What is your earliest memory? Use the space below to tell about it.

I can't wait
to read what
you write!

# CHAPTER 7.5

## MORE ABOUT LEARNING AND MEMORY

There are two types of long-term memory. They are known by several names, none of them easy to remember. Don't worry if your brain has trouble putting the names into long-term storage. Understanding what they are is more important than remembering the names.

The first type of long-term memory is called ***implicit*** *(im-PLISS-it)*. It is also known as the ***procedural memory***, which might be an easier word to remember because it relates to learning "procedures" for doing things like tying shoelaces, riding a bike, brushing your teeth, catching a ball, or playing chords on a guitar. These things require your muscles to make a coordinated pattern of movement. Since coordination is the key, this type of memory is centered in your cerebellum. Looks like the man to the right has some incredible implicit memories! Learning how to do a balancing act is a type of remembering, but not something you can put into words.

The second type of long-term memory is called ***explicit*** *(ex-PLISS-it)*. This is what you normally think of as memory. To further confuse you, however, brain scientists have broken down explicit memories into two sub-categories: ***episodic*** and ***semantic***. Actually, their names make them sound more confusing than they really are. Episodic memory is about an episode in your life, like what you did on vacation last year, or the time your aunt embarrassed you at the family reunion. Semantic memory contains facts that you know but don't remember ever learning. You know that a pencil is used for writing, but you don't remember a specific event where you learned that fact. Math facts might even be semantic. You probably don't remember when you learned that 2+2= 4. You just know it.

In the last chapter we learned that short-term memory can hold only about seven items, with the absolute maximum being nine. Even the smartest people in the world can't hold more than nine things in their short-term memory. How, then, do some people manage to remember long lists, making it appear that their short-term memory can hold far more than nine items? The answer is that the brain does something called ***chunking***. Like the name suggests, chunking has to do with chunks. The short-term memory can hold about seven "chunks" of information. However, those chunks can be anything that the brain considers to be one piece of information. For example, if someone read off this list of numbers, and asked you to repeat them, would you be able to do it? 111, 222, 333, 444, 555, 666, 777, 888, 999 That's a total of 27 numbers. Of course, if would be very easy for you to repeat them, because your brain sees this series of number as about only three pieces of information. The first piece of information is that you start to count; the second piece of information is that you say the number three times, and the third piece of information is that you stop when you get to 9. If you were given a similar string, but the numbers went up to 20, it would still be the same three pieces of information, except that you would stop at 20 instead of 9. Counting is something you don't need to use your short-term memory for. Those counting facts are already in long-term storage. It would not matter how high the list went, it would still be just three pieces of information.

How many chunks of information would you need in order to remember this phone number? 222-1492 You might need only two. The first chunk would be "222," and the second chunk is a famous number: the year that Columbus "sailed the ocean blue." What about this phone number? 531-8642 You need to remember to start with five, then count down by odd numbers, then start with 8 and count down by even numbers. That's well under the seven-chunk limit. When remembering your friends' phone numbers, you probably do chunking with the first three numbers. You are likely to have memorized the extensions in your area and need only to remember which extension it is. For example, in one community, all the extensions might start with "23": 231-, 234-, 235-, 237-, and 238-. If you already know these extensions, learning a new phone number only requires five chunks of information—the extension, and the four numbers following it. Your brain would see "234" as one chunk.

People who can seemingly remember more than seven items in their short-term memory have learned to use chunking very quickly and efficiently. They may be able to instantly join two pieces of information together to make one. For example, given a string of random numbers, they will join two at a time, remembering the list as a string of seven two-digit numbers, instead of fourteen individual numbers. They would also look for any patterns with significance for them: their age, a date in history, a friend's birth date, and so on, which could provide them with an instant mnemonic.

Another trick memory experts can use is tapping into the edge of long-term memory by using very fast repetition. We know that repeating something over and over again will transfer it from short-term to long-term storage. Those neuronal networks are reinforced each time the thought is repeated. With some practice, the experts can use those split seconds in between pieces of new information to reinforce the previous ones. For example, if you quickly repeat the whole string of numbers from the beginning each time a number is added, it might be just enough to get the first part of that string over into long-term territory. Of course, that doesn't mean you would remember those numbers for a long time. Some long-term storage only lasts for a few minutes or hours. If it is not reinforced again during this time, the memory will fade.

Why do some people have better memories than others? Do they have more glial cells in their hippocampus? Do they have more dendrites per neuron, or more efficient synapses? Brain researchers do not know the answer. Some researchers are studying memory savants like Kim Peek and Daniel Tammet to see how their brains process information. (Daniel Tammet memorized the value of $\pi$ (3.1415...) to 20,000 decimal places!) So far, there is no sure answer as to what exactly goes on inside their brains. What we do know is that each one of us has more brain potential than we will ever use. There is nothing more complex and more amazing than a human brain, including yours!

* * * * * * * * * * * * * * * * * * * * * * * * * * * * * * * * * * * * * * * * * * * *

## ACTIVITY 7.5 Learning style survey

What is your favorite way to learn? Do you prefer looking, listening, or doing? Some people are visual learners, some are audio learners, and some are kinetic learners. If you are a visual learner, you like to see and read, and you do well learning from books. If you are an audio learner, you remember things best if you hear them. If you are a kinetic learner, you need to physically move your body in some way to learn well. This survey can help you determine whether you are primarily a visual, audio, or kinetic learner. (Everyone uses all three modes at various times, but most people have a preferred mode.)

For the following questions, choose the best answer. It is important that you not spend too much time answering. The choice that comes to your mind first is the one to mark down.

1) If you were planning a vacation trip, how would you go about it?
   a) Make a check list of everything I need and everything I want to see
   b) Make a few phone calls and talk to someone who has been on a similar trip
   c) Imagine myself actually on the trip and plan according to my mental images

2) Which are you most likely to do while waiting in a line?
   a) Look at posters if there are any, or watch the other people who are in line with me
   b) Talk to the person next to me in line
   c) Tap my foot, pace, or move around in some way

3) You have just entered a museum. What do you do?
    a) Find the information booth and get a map
    b) Go over to a museum staff person and ask advice about where to go first
    c) Strike out in any direction and rely on your sense of direction to guide you

4) Which would you be most likely to do if something exciting happened to you?
    a) Tell someone else then watch their face to see the reaction
    b) Tell someone the whole story of why I am happy, even if it takes a few minutes
    c) Jump up and down, nod my head, make gestures with my ams or hands, etc.

5) Of these options, which do you prefer?
    a) Read a story all by myself
    b) Listen to someone read the story to me out loud
    c) Act out the story in a drama

6) If you are trying to study or concentrate, which do you find most annoying?
    a) Things or people I see in the room out of the corner of my eye
    b) Noises in the room
    c) Feelings inside myself such as hunger, thirst, shoes are too tight, itchy sweater, etc.

7) If you are trying to remember how to spell a word, what do you do?
    a) I would write the word several ways and see which one looks right
    b) I would sound it out or spell it out loud
    c) I would write the word several ways and see which one feels the most familiar

8) If you had to learn how an engine works, which you would rather do?
    a) Read a book and look at diagrams
    b) Have someone explain it to me
    c) Tear one apart and figure out how it works as I go along

9) Of the following subjects, which one are you best at? If more than one, go ahead and circle any that apply.
    a) Art
    b) Music
    c) Sports

10) When you listen to a song, which of these is most true?
    a) I see pictures in my mind. The song reminds me of places I've been or people I know.
    b) I hum along and I know I will be able remember the melody for a quite a while.
    c) The rhythm is the most important part of the song. I would probably tap my foot or dance.

11) Of the following choices, which one is the worst? (Which would be the most difficult for you, personally?)
    a) Going blind
    b) Going deaf
    c) Being paralyzed

12) Of these choices, which book would you prefer?
    a) A book about another country that has a lot of photos and maps
    b) An autobiography where the author tells stories of his or her childhood
    c) An activity book that recommends that you try the things they describe for yourself

13) Of the following options, which one is the worst?
    a) A house with orange countertops, green carpet, brown furniture, and blue wallpaper
    b) A house where the television and stereo are always blaring too loud

c) A house with uncomfortable chairs and a temperature that is always too hot or cold

14) When you were younger, which activity would you have enjoyed the most?
    a) "I spy with my little eye..." or a scavenger hunt where you look for hidden items
    b) Listening to Mother Goose rhymes
    c) Playing "London Bridge" or "Ring Around a Rosy"

15) If you went to a lecture and wanted to remember what the speaker said, what would you do?
    a) Spend most of my time looking at the speaker, but take a few notes
    b) Listen intently, maybe even closing my eyes to concentrate
    c) Write down as many notes as possible, and doodle on the page if there is nothing to write

16) If you could do ONLY ONE of the following to learn a spelling word, which would be your choice?
    a) Look at the word for 30 seconds
    b) Listen to someone spell the word over and over till the 30 seconds are up
    c) Look at the word, then write it as many times as possible in 30 seconds

17) Which of these is easiest for you to remember?
    a) Faces
    b) Names
    c) Dance steps

Now score your results.
Number of a's you circled: ______ Number of b's: ______ Number of c's: _______

If you have more a's than anything else, chances are you're a visual learner. If you have more b's, you are probably an audio learner. If you have more c's, it is likely that you are a kinetic learner. If you have two or more with equal scores, that would suggest that you have two learning modes that are equally strong.

Remember, though, everyone uses all three of these learning modes at least some of the time. It's only a matter of which one suits you the best. (Please note that this survey cannot be used to officially diagnose anyone. It's just a silly survey.) Also, the younger you are, the less helpful surveys like this will be. Every year that you are alive adds to your life experience and your knowledge of yourself.

## ACTIVITY 7.6 Categorize these memories

Put a check on the line that best describes each type of memory. Remember, implicit memory is in the cerebellum and is also called procedural memory. Episodic memory is tied to a certain event and semantic memory is made of facts you know whether or not you remember learning them. A few of these could have more than one right answer, depending on your life experience.

| | Implicit (procedural) | Explicit (episodic) | Explicit (semantic) |
|---|---|---|---|
| The phone number of Carlo's Pizza | ______ | ______ | ______ |
| The feel of my dog's fur | ______ | ______ | ______ |
| 3x3=9 | ______ | ______ | ______ |
| My math lesson from last week | ______ | ______ | ______ |
| How my yard looks in the winter | ______ | ______ | ______ |
| How to tie my shoe | ______ | ______ | ______ |
| How to do a certain dance | ______ | ______ | ______ |
| Where the Hawaiian Islands are located | ______ | ______ | ______ |
| How my bedroom looks right now | ______ | ______ | ______ |
| How to type on the keyboard without looking | ______ | ______ | ______ |
| How to spell the word "brain" | ______ | ______ | ______ |
| Don't mix peppermint with Cola-- yuck! | ______ | ______ | ______ |
| I must be home by 9:00 PM | ______ | ______ | ______ |
| How to zip a zipper | ______ | ______ | ______ |
| The nursery rhyme "Humpty Dumpty" | ______ | ______ | ______ |
| My homework assignment for tomorrow | ______ | ______ | ______ |
| The dog next door bites | ______ | ______ | ______ |
| What I got for my birthday last year | ______ | ______ | ______ |
| The fact that birds can fly | ______ | ______ | ______ |
| How to sign my name | ______ | ______ | ______ |

Have you ever tried peppermint with cola?
Yeah. It really does taste yucky!

### ACTIVITY 7.8 A word puzzle

Fill in the correct answers to these clues, then match the corresponding numbers underneath to the letters in the answer to the question below.

1) This is when your brain puts several small bits of information together to make one larger whole. ___ ___ ___ ___ ___ ___ ___ ___ (11, 3, 15)

2) This type of memory is based in your cerebellum, instead of your cerebrum. ___ ___ ___ ___ ___ ___ ___ ___ (1, 6, 12, 20)

3) This is another name for the type of memory in 2). ___ ___ ___ ___ ___ ___ ___ ___ ___ ___ (10, 16, 18, 5)

4) This type of memory contains specific events that happened to you. ___ ___ ___ ___ ___ ___ ___ ___ (2, 13, 22, 21)

5) This type of memory contains facts that are not linked to a specific memory. ___ ___ ___ ___ ___ ___ ___ ___ (9, 23, 4)

6) This type of memory includes both the categories listed in 4) and 5) ___ ___ ___ ___ ___ ___ ___ ___ (8, 19, 17)

7) The short-term memory can only hold about this many pieces of information. ___ ___ ___ ___ ___ (7, 14)

*QUESTION: What are the two most important factors in determining how well the brain keeps its memory ability as it gets older?*

___ ___ ___ ___ ___ ___    ___ ___ ___ ___ ___ ___ ___ ___
1 2 3 4 5 6    7 8 9 10 11 12 13 14

and ___ ___ ___ ___ ___ ___ ___ ___ ___
15 16 17 18 19 20 21 22 23

Doctors now recommend that elderly people learn new skills.

# CHAPTER 8

## HOW THE BRAIN CONNECTS TO THE BODY

The brain and the spinal cord make up what is called the ***central nervous system***. All the other nerves outside the brain and spinal cord are referred to as the ***peripheral*** *(per-IF-er-ul)* ***nervous system***. Peripheral means "around the outside." The peripheral nerves include your five senses, the nerves that control the movement of your muscles, and some automatic body systems that we will learn about in this chapter.

This chart shows the parts of the nervous system and how they are organized into groups. We've already mentioned the first division, into the central nervous system (CNS) and the peripheral nervous system (PNS). The central nervous system is just the brain and the spinal cord.

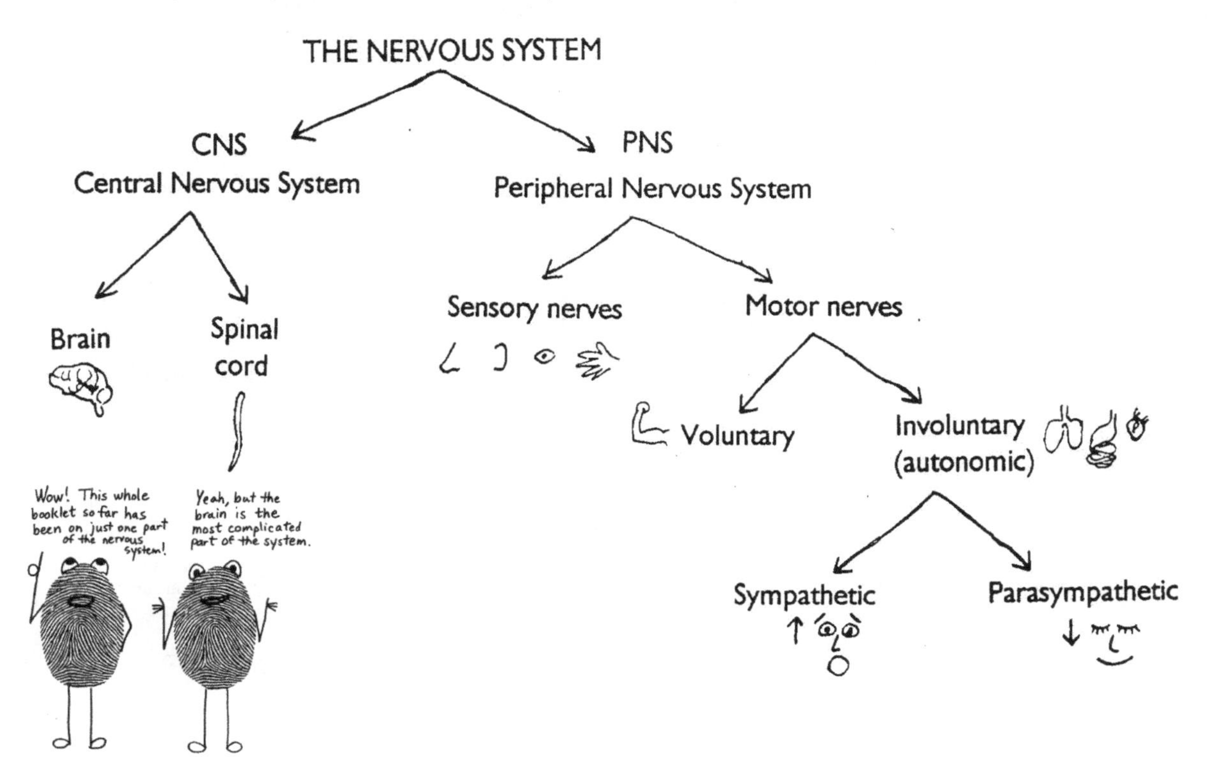


All peripheral nerves are either ***sensory*** nerves or ***motor*** nerves. As the word sensory suggests, the sensory nerves are the ones found in your five senses: sight, hearing, taste, smell, and touch. The motor nerves control muscles, both the muscles that you can move voluntarily, like your arms and legs, as well as the ones that you don't have to think about, like the muscles in your heart and your intestines.

You are well acquainted with your ***voluntary*** muscles because you have direct control over them. You tell your motor cortex to send out signals to the voluntary muscles in your arms, telling them to move a certain way. You are not as familiar with your ***involuntary*** *(autonomic)* muscles because they function automatically—you don't have to think about them. Your heart beats, your diaphragm contracts, your stomach churns, your intestines push food along, the pupils of your eyes get larger and smaller, your saliva glands squeeze out saliva, and your blood vessels tighten or relax to adjust your blood pressure. You aren't even aware you are doing these things because your autonomic system just takes care of it all for you.

The ***autonomic*** part of the nervous system can be divided into two categories: ***sympathetic*** and ***parasympathetic***. The names of these categories might be hard for you to remember because they don't seem to have a whole lot to do with the function of these muscles. From the name "sympathetic," you might guess that these muscles are compassionate and feel sorry for other muscles (maybe ones you used too much the day before), since "sym" means "with" and "pathos" means "suffering." A better definition for our purposes

would be "with great passion or excitement." This is the body system that gears you up when you need to be alert. The sympathetic nerves tell all their muscles to speed up. This makes your heart beat faster and your breathing rate increase. Your adrenalin glands start pouring adrenalin into your bloodstream. You begin to sweat, your mouth goes dry and the pupils in your eyes open up wider. Your muscles feel all tingly and ready for action because of the adrenalin. This is exactly what your body needs for some situations. Any time there is an emergency, your body flips the switch and turns on the sympathetic system. You are ready to go!

After the emergency is over, you body wants to return to normal. You can't be on "red alert" all the time. The system that turns off the sympathetic system is called the parasympathetic system. "Para" means "against." The parasympathetic system slows down the heart and breathing rates and calms the muscles, getting the body back to normal. The pupils in your eyes get smaller, you begin to salivate again, and you stop sweating. The parasympathetic system also helps you to recover after exercising.

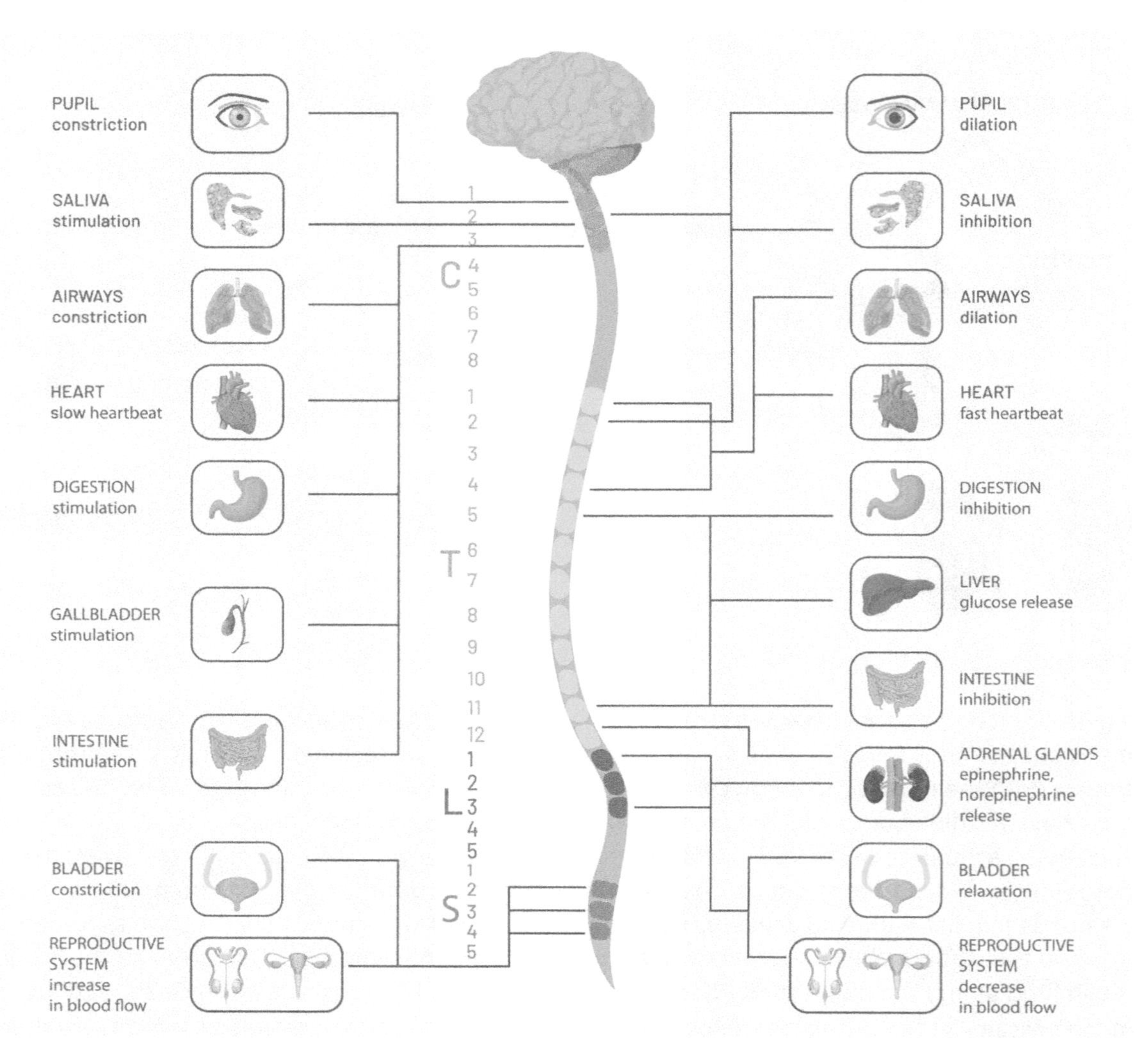


The colored dots on the spine present groups of vertebrae that we give certain names. (There aren't really any blank spaces between these groups; the artist chose to space them out so all the pictures on the sides would fit on the page.) "C" stands for "cervical" (neck), "T" is for "thoracic" *(thor-RASS-ik)* (chest), "L" is for "lumbar" (low back), and "S" is for "sacral" (the end of the spine). The lines represent nerve fibers that come out of the vertebrae at these places and go to the organ(s) indicated.

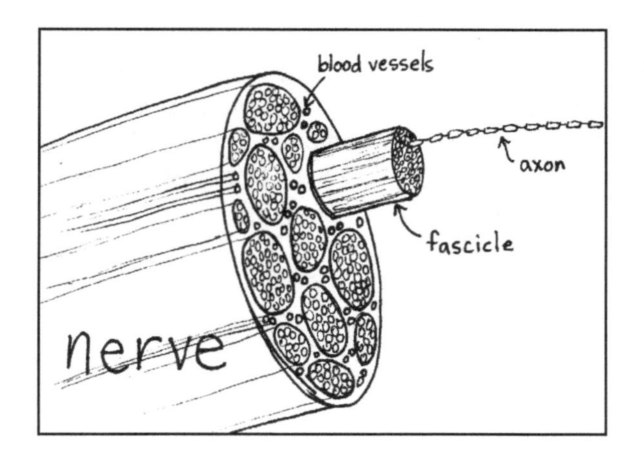


What do these nerves look like? What is a nerve fiber? The diagram on the left shows that a nerve is made of bundles of bundles of bundles. The smallest units are the individual axons of the neurons. The axons are bundled into long units called ***fascicles*** *(FASS-ik-uhls)*. The fascicles are then bundled into the units we call ***nerves***. Each bundle, both the fascicles and the nerves, are covered with a layer of connective tissue, (the same stuff that you find in ligaments and tendons).

All the axons in a fascicle belong to nerves that have similar functions; they might all be headed to your right hand, or to the smooth muscles in your intestines.

Now, let's take a look at a slice, or cross section, of the spinal cord. (This picture does not show the bone [vertebra] around the spinal cord.) The inside of the spinal cord is made of gray matter and white matter, just like the brain. In the spine, however, the pattern is reversed, with the white matter on the outside and the gray matter on the inside. (Remember that gray matter is mostly cell bodies and white matter is mostly axons.) The hole in the center of the cord, the ***central canal***, is an extension of the ventricles (fluid-filled spaces) in the brain. The canal runs from the bottom of the brain to the bottom of the spine.

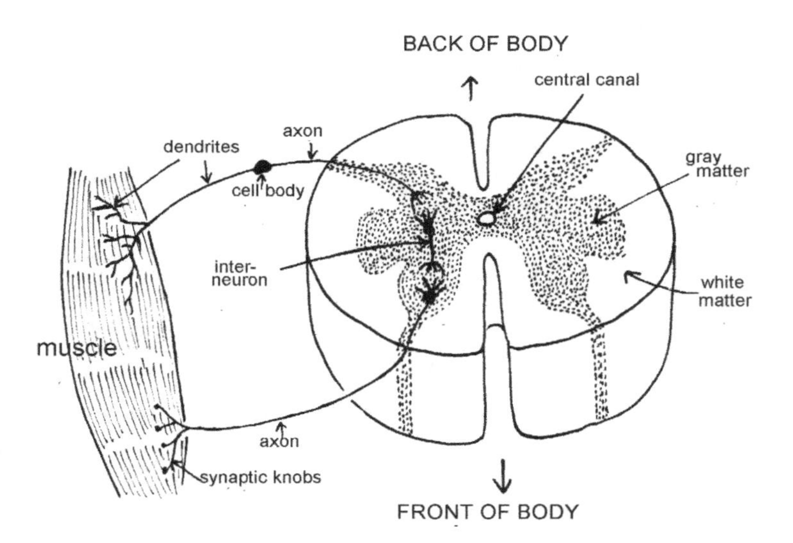


The neurons in this diagram are drawn much larger than they should be, so that you don't need a microscope to see them. The muscle represents any muscle of the body. Notice the synaptic knobs and the dendrites touching the muscle. The knobs are transferring the electric signal to the muscles, and the dendrites are picking up information to send back to the brain. "Inter-neurons" inside the gray matter connect neurons to other neurons, and carry the signal up to the brain. Notice that the neurons carrying signals out of the spine are located on the front side, and the neurons carrying signals into the spine are located on the back side. This holds true all the way up and down the spine.

The ***vertebrae*** protect the spinal cord and prevent the neurons from being damaged. The holes in the vertebrae (shown by **red dot** in vertebra below) line up so the spinal cord can go down through them. There are small gaps between the vertebrae where bundles of nerves can get out (shown in green). There is padding in between the vertebrae bones so they don't rub together. It's amazing how much your back can twist and flex and bend without hurting the nerves!

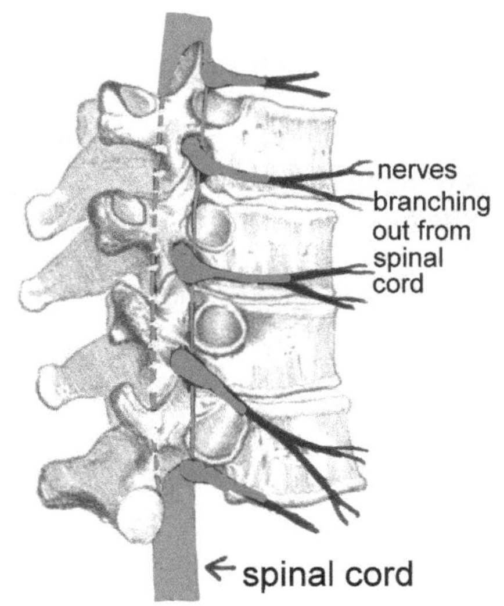


Here's a drawing of a vertebra that Andreas Vesalius did way back in the 1500's. (Remember Vesalius from chapter 1 ?!)

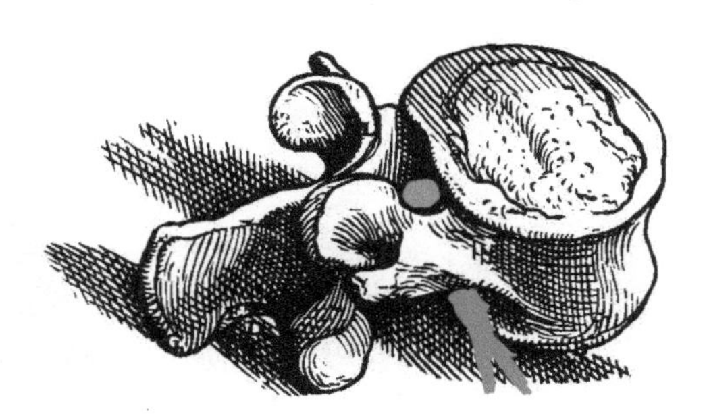

A very important safety feature programmed into your peripheral nervous system is the ***reflex***. A reflex allows your body to react to dangerous situations with lightning speed. Every millisecond counts when you do something like accidentally touch a hot stove, so your peripheral nerves don't even take the time to send a message to the brain before they react. It's much faster to go just to the spinal cord and back. The peripheral nerves take care of the situation right away, then send a signal to your brain to notify you of what just happened.

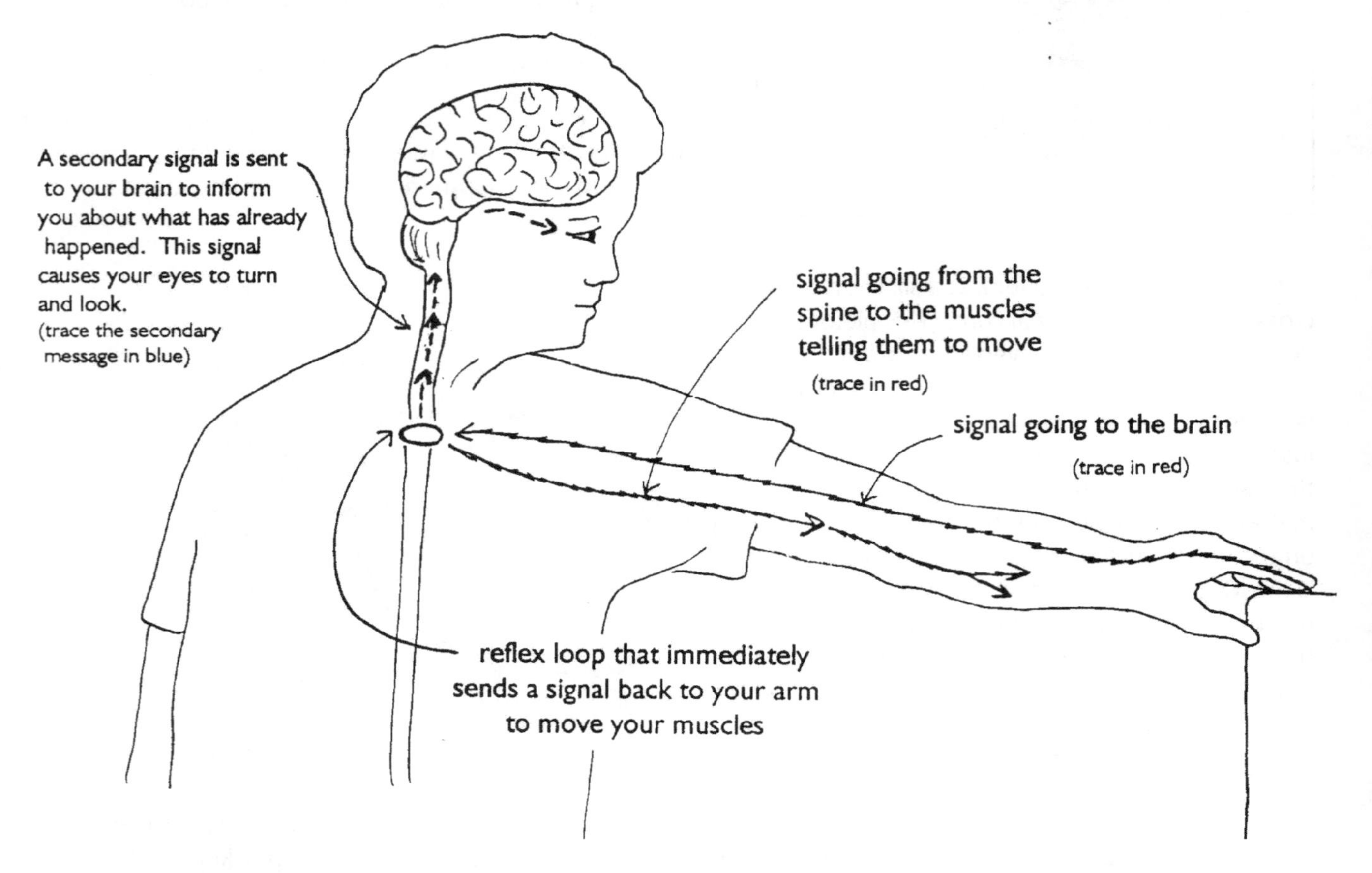
A secondary signal is sent to your brain to inform you about what has already happened. This signal causes your eyes to turn and look.
(trace the secondary message in blue)
signal going from the spine to the muscles telling them to move
(trace in red)
signal going to the brain
(trace in red)
reflex loop that immediately sends a signal back to your arm to move your muscles

Other types of reflexes include blinking, sneezing, gagging, vomiting, yawning, grasping and scratching. The reflex in your knee that doctors sometimes check is called the *patellar* reflex (the patella is the name of your "knee cap"). When the doctor hits the patellar tendon under the knee cap, this stretches the tendon a tiny bit. The tendon pulls on the muscles in your leg, which then send a signal to the reflex area of your spinal cord. The spinal cord sends an instant signal back telling the muscles in the leg to contract a little bit, to compensate for being stretched. This tightening results in your lower leg "kicking" (just a very small amount). Sometimes this reflex is called the *knee-jerk response*.

A reflex you use all the time without knowing it is the ***pupillary reflex***. The pupil in your eye has a muscle around it so that it can make the hole larger or smaller, depending on how much light is available. In low light the pupil opens up to let in as much light as possible. In bright light, the pupil shrinks and becomes very small because too much light will damage the retina at the back of the eye. Your pupils are constantly adjusting to levels of light all day long as the light around you changes.

In bright light
In dim light

## ACTIVITY 8.1 Reflex word search

Use these clues to find reflexes in the word search puzzle. The words in the puzzle may go up, down, diagonally, or backwards.

1) This reflex keeps your lungs and bronchial tubes clear.
2) This reflex brings things up that were not supposed to go down.
3) This reflex is a palindrome: it spells the same backwards and forwards.
4) This reflex is designed to relieve an itchy feeling.
5) This reflex keeps your nose clear of dust or other irritating substances.
6) This reflex is what you do if you suddenly feel like you are falling.
7) This reflex keeps your eyes safe.
8) This reflex is "contagious" and if you see someone else do it, you might do it, too!
9) This reflex is the main one doctors check.

| | | | | | | | | | | | |
|---|---|---|---|---|---|---|---|---|---|---|---|
| C | F | M | Y | H | L | T | H | B | L | Y | M |
| E | Z | E | E | N | S | W | C | L | R | A | R |
| N | D | O | R | L | N | V | T | I | R | W | O |
| B | X | G | A | G | P | S | A | R | G | N | P |
| I | L | S | H | J | L | S | R | M | F | T | F |
| J | T | I | M | O | V | I | C | O | U | G | H |
| O | R | W | N | T | U | D | S | P | L | R | H |
| M | F | S | L | K | H | O | Z | N | G | D | L |
| E | S | P | A | T | E | L | L | A | R | E | Y |

## ACTIVITY 8.2 Watch your pupillary reflex

Look into a mirror. Notice the size of your pupils. Now put your hands over your eyes to give them total darkness. Don't press on your eyeballs. After about 30 seconds or so, uncover your eyes and look into the mirror right away. Watch your pupils. They should go from large to small very quickly. You can repeat this experiment as many times as you want to. If you have someone to be your partner, you can watch each other's pupillary reflexes.

## ACTIVITY 8.3 True or false questions

Put a T or an F beside each statement. Use the diagram on page 67 to help you.

1) All motor nerves are peripheral nerves. ____

2) Both voluntary and involuntary muscles are controlled by motor nerves. ____

3) The peripheral system is part of the central nervous system. _____

4) The sensory nerves are part of the autonomic (involuntary) system. ____

5) Both sympathetic and parasympathetic nerves are part of the involuntary system. ____

6) The brain and the spinal cord are both part of the central nervous system. ____

7) Sympathetic nerves are part of the peripheral system. ____

8) All motor nerves are peripheral nerves, but not all peripheral nerves are motor nerves. ___

9) All sympathetic nerves are involuntary, but not all involuntary nerves are sympathetic. ___

10) Both sympathetic and parasympathetic systems are motor nerves. ___

**ACTIVITY 8.4 Write about a time your sympathetic system turned on suddenly**

Write about a time when your sympathetic system turned on very quickly. Did a large dog frighten you? Did you get scared during a movie? How did you feel, and how long did it take to feel okay again?

## CHAPTER 8.5

### MORE ABOUT HOW THE BRAIN CONNECTS TO THE BODY

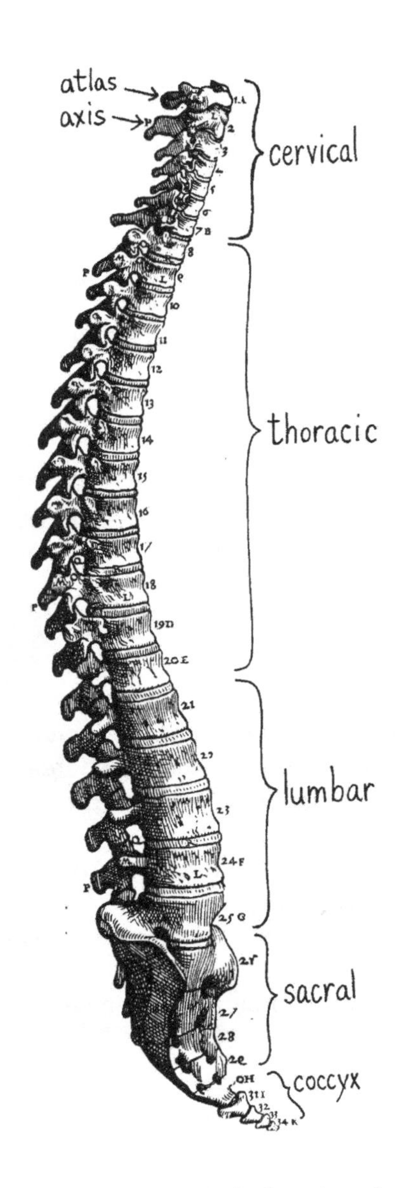


The spinal column is important to the brain because it protects all the nerves that connect the brain to the body. Most people have 33 vertebrae, although rarely someone is born with 32 or 34. It's not harmful to have more or less than 33, just a little unusual. Most people who have 32 or 34 are unaware of it. Rarely do doctors think to count vertebrae.

The top seven vertebrae, which make up the neck, are called the ***cervical*** *(SER-vik-ul)* vertebrae. (FUN FACT: Almost all mammals have seven cervical vertebrae, no matter how long or short their neck is. Even a giraffe has only seven.) The top two cervicals, right under the skull, have special names. The very top one is called the ***atlas*** because it bears the weight of the skull sitting on top of it, just like the mythological character Atlas who was said to carry the Earth on his shoulders. Under the atlas is the ***axis***. It allows the skull to rotate on top of the neck. The other cervical vertebrae are just known by numbers: C3, C4, C5, C6, and C7.

Atlas carried the world on his shoulders.

The next section of vertebrae are ***thoracic*** *(thor-RASS-ik)*. The numbers start over again, so these are numbered T1 to T12. (They are actually vertebrae #8- #20.) From the thoracic vertebrae come all the nerves that go to the heart, lungs, stomach, and other organs in your chest. The ***lumbar*** vertebrae are numbered L1 to L5. Their nerves go out to the lower back and legs. Below the lumbar, are the ***sacral*** vertebrae. On the very bottom, the last few vertebrae are permanently stuck together to form the ***coccyx*** *(COK-sicks).*

Notice the little pads between the vertebrae. They are called ***discs***. If a disc slips out of place it can be very painful and might require surgery to put it back. Fortunately, the discs are anchored securely and rarely slip out.

If the vertebrae get out of line, they can pinch the nerves that stick out of them. Doctors who have been trained to push the back into place are called ***chiropractors***. "Chiro" means "hand." Chiropractors push and pull the spine with their hands. Sometimes they use other treatments, as well, such as massage or physical therapy exercises.)

Bundles of nerves go from the spinal cord out to all parts of the body. As we have mentioned previously, some of these nerves are sensory nerves, which are busy sending signals back to the brain telling it what is going on, both in your body and in the environment around you. Each sense has a special kind of nerve ending. The nerve endings at the back of your eyes (in an area called the retina) are designed to sense photons of light. The nerve endings that sense light can't sense pain, heat, sound or taste. The nerve endings in the nose can sense only odor, nothing else. Those in the ear can sense only sound. Probably the most interesting-looking nerve endings are found in the skin. There are five different types, each one able to sense only one thing.

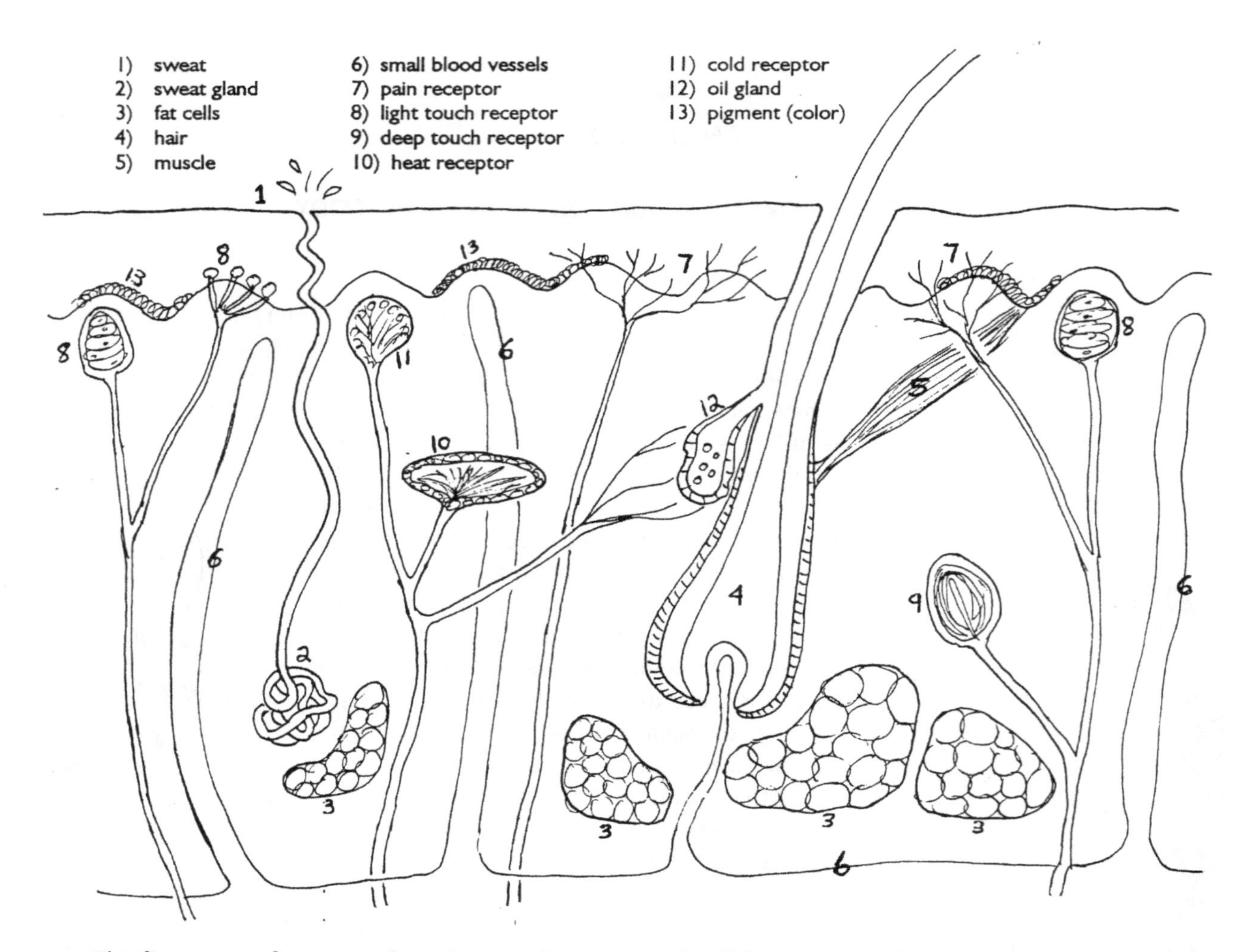


The five types of nerve endings in your skin sense ***pain, light pressure, deep pressure, heat, and cold.*** Notice that the nerve endings for pain look unprotected (ready to "scream" if anything touches them), whereas the pressure receptors have padding around them. There are also lots of other things in the skin, as you can see.

The sensory nerve cells in the eye are called ***rods and cones***. The rods sense light, but not color. The cones can sense color, but need sufficient light to function. When the light is dim, your cones don't work, but your rods can see shades of gray. Thanks to your rods, you can see well enough at night to be able to make your way to the bathroom without bumping into the walls.

Intermediate cells transfer light signals into electrical signals that are sent to the brain. The main nerve that takes the information to the brain is the ***optic nerve***.

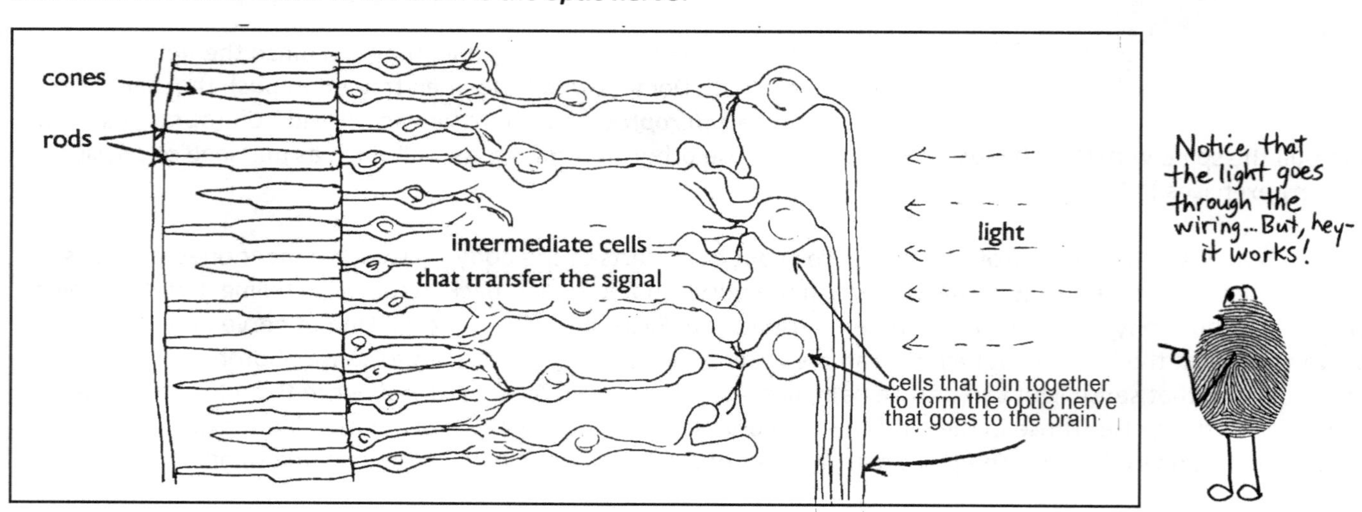

Diagrams of the nerve endings in the nose and tongue don't look quite so complicated:

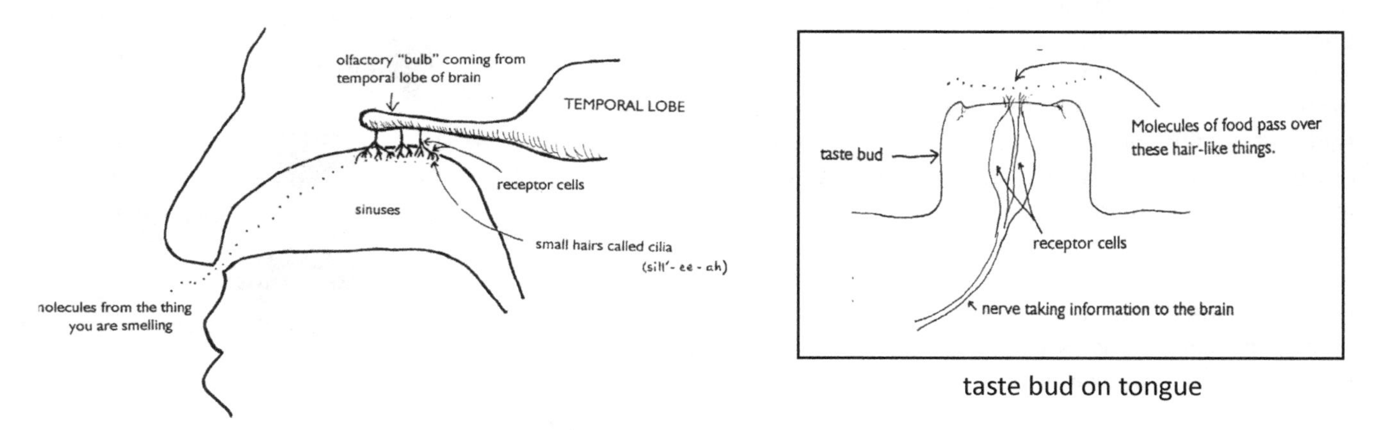


taste bud on tongue

The nerve endings in both the nose and mouth are designed so that certain types of molecules will stick to certain of the receptor cells, causing those cells to fire off a signal to the brain. There are extremely small hairs (not the nose hairs you can see in your nose) in the sinuses that aid the nerve endings in picking up signals.

There are also microscopic hairs in the ear that help the auditory nerve endings. Deep inside the ear, behind the ear drum, in the ***cochlea*** *(KO-klee-ah)*, these microscopic hairs help to transfer sound waves to the nerve endings. The pressure of the sound waves moves the hairs, which "tickle" the nerve endings, causing them to send electrical signals to the brain. The nerve that carries sound information to the brain is the ***auditory nerve***.

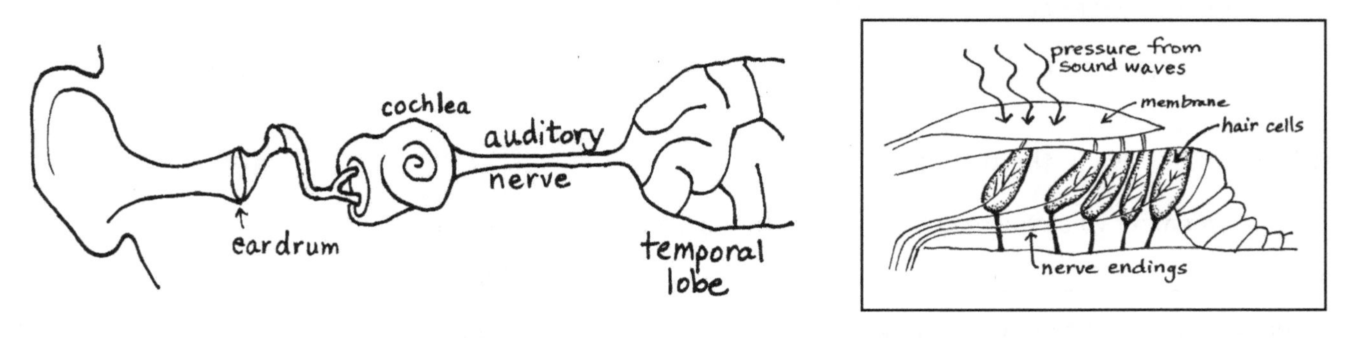


Here's a bizarre, but interesting, question: What would happen if the neurons taking in sound from your ears went to the vision center in your occipital lobe, instead to the sound center in your temporal lobe? Would you see sounds? What if the nerve endings in your eyes were wired to your temporal lobe? Would you hear colors? Yes, you probably would!

It was once thought that "mixed up wiring" is what causes an unusual condition known as ***synesthesia*** *(SIN-es-THEE-zhee-uh)*. People with this condition seem to have sensory input going to the wrong parts of the brain, as they report experiences such as seeing sounds, hearing or smelling colors, or tasting words. A famous painter named Kandinsky always said that he could "hear the colors in his paintbox hissing."

Recent research has revealed that it is unlikely that sensory nerves are actually wired to the wrong lobes. Certain drugs and medicines can cause temporary synesthesia that goes away within a few days. The wiring in your brain can't change that fast. So a new hypothesis was needed to explain what was going on. We know that each brain part is connected to every other brain part in some way. Some of these connections are direct and strong (such as between the frontal lobe and the motor cortex) but others are supposed to be indirect and not very strong. Synesthesia might occur when connections that are supposed to be weak become too strong. Another line of evidence supporting this hypothesis comes from the realm of language, and our ability to use analogies. We say that cheese tastes "sharp." The color red evokes feelings of danger or excitement. Words like "oval" and "blue" sound soft, whereas words like "cut" and "people" sound hard. To use analogies, brain parts have to share a little bit of information. This type of weak connection is normal. It is possible that in synesthesia, these normal connections are just too strong.

There are different types of synesthesia. For some people, words or numbers to appear colored, even though they are printed in black and white. One person reported that the number five is bright red, whereas the number two is yellow. Others say that certain words give them taste sensations. One person said that reading the name "Lori" tasted like a pencil eraser, whereas reading the name "Laurie" tasted like lemon. The artist Kandinsky saw music as colors and shapes. His paintings were often pictures of what he saw when he listened to a particular piece of music. For other synesthetes, tastes evoke colors. One man reports that chicken tastes light blue. When he cooks, he matches colors as well as tastes.

"Yellow-Red-Blue" by Wassily Kandinsky, 1925

* * * * * * * * * * * * * * * * * * * * * * * * * * * * * * * * * * * * * * * * * * * * * * * * * * * * * * * * *

### ACTIVITY 8.6 Look at some paintings by Kandinsky

Use an Internet search engine set on "images" and use key words "Kandinsky paintings."

### ACTIVITY 8.7 Meet some people with synesthesia

There are some fascinating interviews with synesthetes posted on the YouTube channel.

### ACTIVITY 8.8 Fill in the blanks

Can you remember what you read? If you have trouble, go back to the text and find the answers.

1) The __________________ nerve carries signals from the ear to the brain.

2) The vertebrae in the neck are called the _________________ vertebrae.

3) Rods and cones are found at the back of the _______.

4) The skin has two types of touch receptors: ___________ touch and __________ touch. (look at diagram)

5) The nerve that runs from the eye to the brain is the _____________ nerve.

6) The olfactory bulb takes information from the ________ to the ____________ lobe of the brain.

7) The vertebrae in the middle of the spine are called the ______________ vertebrae.

8) The first vertebra right under your skill is called the _________.

9) The vertebrae that are numbered L1 to L5 are called the _____________ vertebrae.

10) The nerve endings in the skin that are unprotected (look like dendrites) are the ones that sense _________.

11) Signals from the inner ear go to the ______________ lobe of the brain.

12) The cells in your eye that sense color are the ___________.

13) The inner ear part that looks like a snail is called the ____________.

14)

15)

# CHAPTER 9

## SLEEP

What happens when we sleep? Why do we need to sleep? Why do we dream? These questions are still being investigated by brain researchers, using "sleep labs." Volunteers agree to spend a whole night in a sleep lab, with wires taped to their head. The wires pick up the electricity their brain produces as they sleep, and a machine displays this electricity as wavy lines on a chart. (More recently, the wavy lines are on a computer screen, not a paper chart.) The name of this machine is the ***electroencephalograph*** (ee-LECT-ro-en-CEFF-al-o-graf) abbreviated as either ***EEG*** or ***ECG***. (*Electro*= electricity, *en*= in, *cephalo*= head, *graph*= picture.)

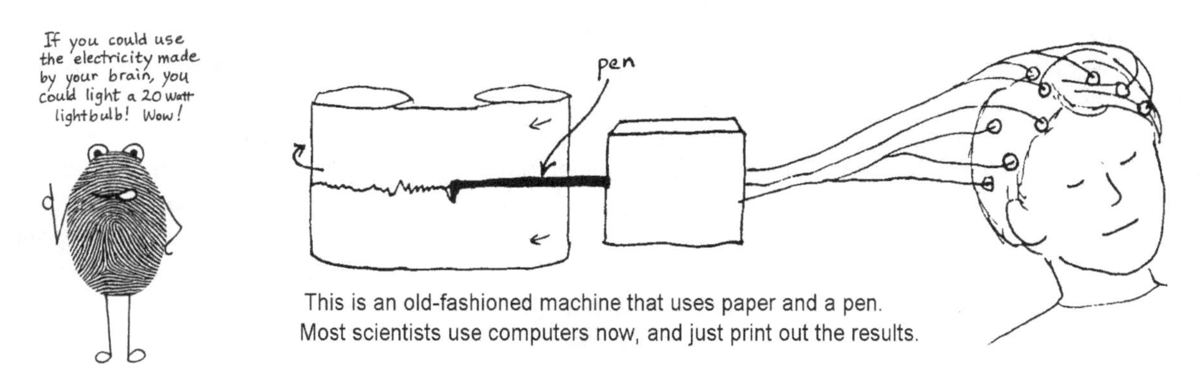

If you could use the electricity made by your brain, you could light a 20 Watt lightbulb! Wow!
pen

This is an old-fashioned machine that uses paper and a pen.
Most scientists use computers now, and just print out the results.

In the morning, the researchers can look at the complete record of what the brain did all night. The squiggles on the chart might look something like this:

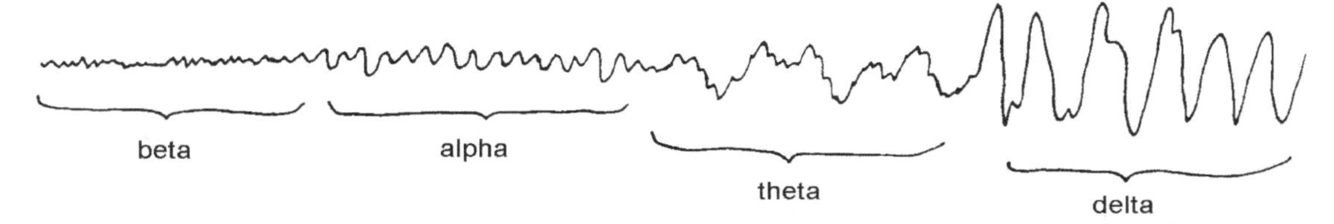


Looking carefully at the squiggly patterns, scientists have determined that brains make four basic types of wave patterns. They decided to name these four patterns after four letters in the Greek alphabet: ***alpha, beta, theta, delta***. In English, it might be like calling them A, B, C, D, (except that theta represents "th," not "c"). When you are wide awake your brain makes ***beta waves***. They are the smallest and shortest waves. As you start to get drowsy, your brain begins to make ***alpha waves***. You also make alpha waves if you are peacefully daydreaming or doing something very relaxing. Gradually, those alpha waves begin to turn into ***theta waves***. This is when you are actually dropping off to sleep. Finally, you go into deep sleep and produce ***delta waves***. Your heart rate and blood pressure drop. You are fast asleep. You might think that this is the end of the story—that you make delta waves until you wake up. But, no, your brain has just begun!

After you have been asleep for about an hour or so, your brain starts making waves that are similar to the ones you make when you are awake. During this time, your eyes move back and forth rapidly, so this stage of sleep is known as "Rapid Eye Movement," ***REM***. It is during REM that you dream. After a period of REM, you go back into deep delta waves for a while. You alternate periods of REM and delta waves until you wake up. A typical night may have four or five periods of REM.

There are many theories about why we dream, but no one knows for sure. Some scientists say that our brain is doing something similar to defragging a computer—sorting and storing all the data it has taken in during the day. Other scientists think our brain is just firing off neuronal networks at random, and our poor frontal lobe has to try to make sense of all the thoughts and tie them into some kind of story line. Some psychologists theorize that our dreams are how we deal with fears and anxieties. All of these could be true. Whatever the purpose of dreams, it is definitely true that your brain uses bits and pieces of memories stored in the neocortex. If you decide to watch a scary movie, just remember that it will become a permanent part of the available material your brain has to work with when it creates dreams!

We all know that feeling of "it's time to go to bed." Our bodies have a natural rhythm of waking and sleeping. This rhythm is known as the ***circadian** (sir-CADE-ee-un) **rhythm***. "Circa" means "cycle" and "dia" means "day." An interesting study was done in Germany to determine what the natural human circadian rhythm would be if there were no sunrise, no alarm clocks, and no one to tell us to go to bed. Volunteers agreed to spend an entire month underground, with no clocks and no sunlight. They had lamps they could turn on when they wanted light. They ate and slept whenever they felt like it. All during this time, they wore brain wave monitors that kept track of when and how much they slept. The results of the study indicated that, if left to our own devices, humans would prefer to have a daily cycle of 24.6 hours, not 24 hours. We really would go to bed a little later every night if we could get away with it! Life brings us extra tiring days every once in a while, which help tire us early and reset our circadian clocks.

Research with rats has suggested that the hypothalamus plays an important part in circadian rhythm. It is linked to both the optic nerve and the pons. The pons is the actual alarm clock that brings you to consciousness, but the hypothalamus helps to determine how many hours of sleep you need, and whether you are a "night owl" or a morning person. The link to the optic nerve is to help your brain be awake when it is light and asleep when it is dark. We also learned about the pineal gland in a previous chapter. This gland is connected to special cells at the back of your eye. The pineal gland is most active when it is dark, producing the hormone ***melatonin***, which helps us to sleep. Young children produce twenty times as much melatonin as adults do, which is why babies "sleep like babies."

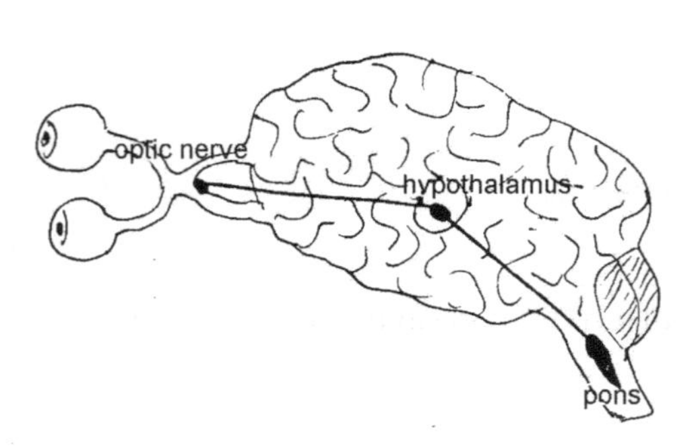


* * * * * * * * * * * * * * * * * * * * * * * * * * * * * * * * * * * * * * * * * * * * * * * * * * * * *

**ACTIVITY 9.1 What do you dream about? (commonly reported dream themes)**

Put a number next to each item, according to how often you dream about it.
0= don't ever remember dreaming out it, 1= dream about it sometimes, 2= dream about it often

___ water (including streams, lakes, etc.)
___ scary encounters with animals
___ being too hot or too cold
___ things I would like to do
___ flying
___ being chased or hunted down
___ traveling to another country
___ trying to find a bathroom
___ meeting someone famous
___ shopping
___ things I did that day

___ natural disasters (tornadoes, etc.)
___ friendly encounters with animals
___ romance
___ performing in front of people
___ falling
___ doing something bad I would never really do
___ scary things happening to my family
___ being someplace without appropriate clothing on
___ music
___ being in school, or taking a test
___ other: ______________________________

SUGGESTION: Compare your results with friends or other members of your family.

# CHAPTER 9.5

## MORE ABOUT SLEEP

Let's take a closer look at the waves the brain makes while sleeping. There are five stages of sleep.

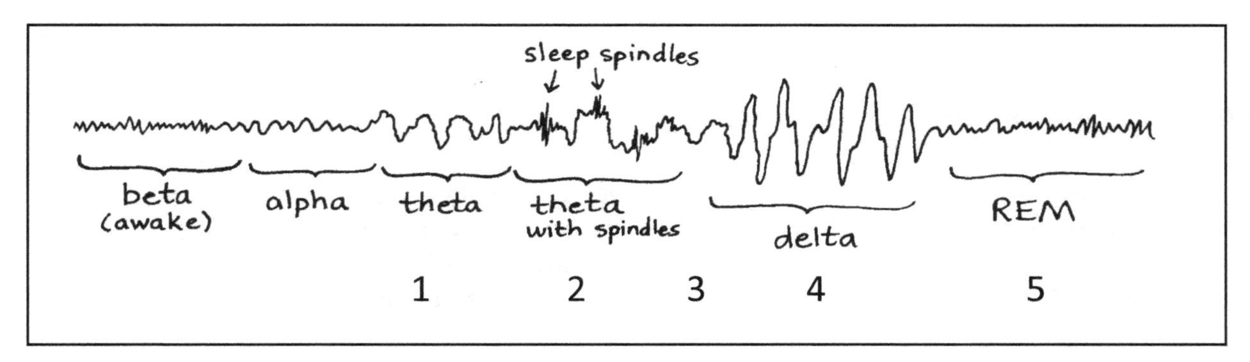


**Stage 1** is made of theta waves and occurs when you are falling asleep. During stage 1 you are vaguely aware of sounds around you, and your muscles may twitch. (On rare occasions, one of your muscle twitches might be strong enough to jolt you awake again.) **Stage 2** begins when your theta brain waves start making patterns called ***spindles***. No one really knows what these spindles do, but we know that during stage 2 you are no longer aware of your environment; you become completely unconscious. **Stage 3** begins when your brain starts making delta waves. Stage 3 doesn't last very long and functions mostly as a transition into **stage 4**. Your deepest sleep occurs during stage 4. This is when you are the most difficult to wake up. It is also the stage during which events such as sleep-walking, sleep-talking, bed wetting, and night terrors occur.

**Stage 5** is REM sleep, when your eyes move rapidly back and forth (Rapid Eye Movement) and your brain waves resemble the type you make when you are awake. Your frontal lobe and motor cortex are busy sending signals during REM sleep. You would actually get up and do what you are dreaming about if it were not for a special function of the brain stem in which it stops the production of one of the neurotransmitters necessary for the muscles to receive the signals. The electrical signals come to the synapses and stop. There are no chemicals there to transfer the signals across the gap. Therefore, the muscles don't ever receive the messages.

REM sleep is when most of your dreaming occurs. Scientists used to think that dreaming occurred exclusively during REM, but sleep labs now report that they have recorded periods of dreaming during all stages of sleep. During the last three hours of your night, you spend more time in REM sleep than any other stage. It is during this time that you have the dreams you will remember when you wake up. People who claim they don't dream just don't remember what they dreamed. If they were tested in a sleep lab, their brains would show the same pattern of REM that everyone else's does.

During the night, your brain goes through the stages of sleep four or five times. If you wrote down all the stages (using the numbers in the diagram above), the series of numbers might be something like this:

1 2345 2345 2345 2345 2345 1

WHAT IS A NIGHT TERROR?

Night terrors aren't the same as nightmares. Nightmares are scary dreams produced during REM sleep. Night terrors occur during non-REM sleep, usually just an hour or two after falling asleep. The person may look awake, but will be seeing and hearing things that aren't real. Often, what they see and hear is frightening to them, and they may yell or cry. It can be distressing to watch someone having a night terror. You won't be able to wake them up out of it, though, and trying to wake them up will only make things worse. The best thing to do is to get them to lie back down and go back to sleep. In the morning they won't remember a thing about it!

If you are awakened too many times during the night and you can't complete enough "2345" cycles, you will probably feel drowsy and tired the next day. If the brain is deprived of REM time, it will go into REM more quickly the next night.

A condition called ***sleep apnea*** occurs when the airways (usually sinuses and/or throat) close up and the brain has to wake up the body to correct the problem. As soon as the person is a awake, the airways go back to normal, but when they fall back asleep, it happens again. To correct this, doctors often recommend using a special breathing apparatus that keeps the airways open.

* * * * * * * * * * * * * * * * * * * * * * * * * * * * * * * * * * * * * * * * * * * * * * * * * * * * * * * * * * * * * *

## ACTIVITY 9.2 Do some REM observation

You may have already observed this, but if you haven't, it's a fun thing to do. Find someone who is taking a nap and watch their eyelids. If you are lucky you'll catch them in the middle of some REM sleep and you will see the eyeballs moving around under the eyelids. If they aren't in REM, wait a while and then come back and try again. (This works with dogs, too. Watch their eyes flutter around. Sometimes they softly bark or making running motions with their feet, too.)

## ACTIVITY 9.3 Read a poem written while the poet was asleep

Samuel Taylor Coleridge was a poet who lived in England in the late 1700s. In the summer of 1797 he was in poor health and retired to a lonely farmhouse to get some rest. One day he took some medicine which made him drowsy, and he fell asleep in his chair just as he was reading the following lines in a book: "Here the Khan, Kubla, commanded a palace to be built and a stately garden thereunto. And thus, ten miles of fertile ground were enclosed within a wall." Coleridge slept for about three hours, during which time he had an incredible dream, so vivid and real that he almost could not believe his brain had been capable of making it up. While he slept, the lines of a poem came to him as if they were already composed and all he had to do was write them down. Upon wakening, he wrote down as much as he could remember. After a few minutes of writing, someone came and called him away from his desk. Upon his return to his desk he found that his memory of the dream had faded considerably. He did his best to patch together what he could in order to finish the poem.

Before you start reading, you will need to know what some of the words mean. Here is a list of the words that could possibly be unfamiliar to you.

*sinuous: winding, wavy like a serpent*
*rill: a very small stream*
*athwart: across, in opposition to*
*cedarn: made of cedar*
*waning: decreasing in size; diminishing*
*chaff: lightweight debris blown off wheat seeds*
*thresher: person who beats (threshes) wheat seeds until the chaff separates and blows off*
*flail: tool used for threshing wheat*
*meander: to follow a wandering path*
*tumult: uproar or confusion*
*prophesy: to predict the future*
*dulcimer: a stringed musical instrument*
*Abyssinian: from Abyssinia, in Africa*

## *"Kubla Khan"* ***by Samuel Taylor Coleridge***

NOTE: You can listen to a professional reader recite this poem on the YouTube channel. (If it's not there, just search using the title and author.)

*In Xanadu did Kubla Khan*
*A stately pleasure-dome decree :*
*Where Alph, the sacred river, ran*
*Through caverns measureless to man*
*Down to a sunless sea.*
*So twice five miles of fertile ground*
*With walls and towers were girdled round :*
*And there were gardens bright with sinuous rills,*
*Where blossomed many an incense-bearing tree ;*
*And here were forests ancient as the hills,*
*Enfolding sunny spots of greenery.*
*But oh ! that deep romantic chasm which slanted*
*Down the green hill athwart a cedarn cover !*
*A savage place ! as holy and enchanted*
*As e'er beneath a waning moon was haunted*
*By woman wailing for her demon-lover !*
*And from this chasm, with ceaseless turmoil seething,*
*As if this earth in fast thick pants were breathing,*
*A mighty fountain momently was forced :*
*Amid whose swift half-intermitted burst*
*Huge fragments vaulted like rebounding hail,*
*Or chaffy grain beneath the thresher's flail :*
*And 'mid these dancing rocks at once and ever*
*It flung up momently the sacred river.*
*Five miles meandering with a mazy motion*
*Through wood and dale the sacred river ran,*
*Then reached the caverns measureless to man,*
*And sank in tumult to a lifeless ocean :*
*And 'mid this tumult Kubla heard from far*
*Ancestral voices prophesying war !*
*The shadow of the dome of pleasure*
*Floated midway on the waves ;*
*Where was heard the mingled measure*
*From the fountain and the caves.*
*It was a miracle of rare device,*
*A sunny pleasure-dome with caves of ice !*
*A damsel with a dulcimer*
*In a vision once I saw :*
*It was an Abyssinian maid,*
*And on her dulcimer she played,*
*Singing of Mount Abora.*
*Could I revive within me*
*Her symphony and song,*
*To such a deep delight 'twould win me,*
*That with music loud and long,*
*I would build that dome in air,*
*That sunny dome ! those caves of ice !*
*And all who heard should see them there,*
*And all should cry, Beware ! Beware !*
*His flashing eyes, his floating hair !*
*Weave a circle round him thrice,*
*And close your eyes with holy dread,*
*For he on honey-dew hath fed,*
*And drunk the milk of Paradise.*

## ACTIVITY 9.4 Read about two scientific discoveries made while the scientists were asleep.

A scientist named August Kekule *(keh-kyu-lay)* was a chemist who lived in the late 1800s. He had been working on the problem of how six carbon atoms and six hydrogen atoms could fit together to make the chemical compound "benzene." The rules of chemistry said that every carbon atom had to have four bonds, but hydrogen could only have one bond. Kekule experimented with every possible mathematical combination that could be made with these atoms, but nothing worked. It had been proven that there were six carbons and six hydrogens in benzene, but no one could figure out what the molecule looked like. Then, one night, Kekule fell asleep while thinking about the benzene molecule. He dreamed that he saw the chain of carbon atoms bend around and touch at the ends, to form a circle. When he woke up, he remembered the dream and realized that his brain had solved the mystery while he was asleep. The benzene molecule was not straight, as everyone supposed—it must be a ring! The ring structure was later proved to be correct.

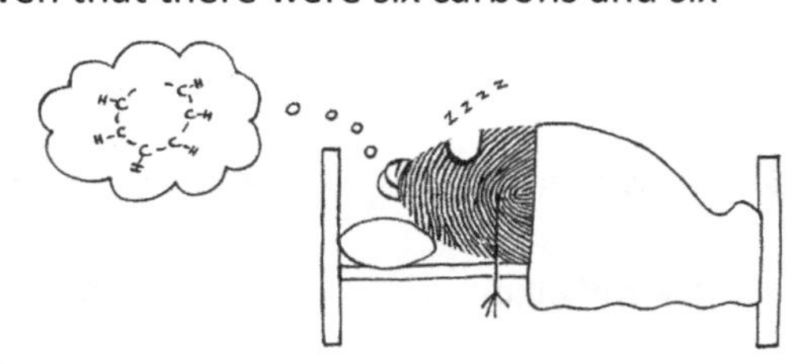

Dmitri Mendeleev is know today as the inventor of the Periodic Table of chemical elements. The table is a rectangular chart with all the known elements listed in columns and rows. In Dmitri's day, only 63 elements were known. It was recognized that there were groups of three elements, called triads, that had very similar characteristics, but that was the extent of any organized patters. Dmitri make paper cards with the name of an element written on each one. He spent many days shuffling the cards around, looking for a pattern. One evening, after working relentlessly for two days straight, he fell asleep over his cards. He dreamed that the cards arranged themselves into a rectangle, with certain cards at the beginning of each line. When he woke up he tried the pattern he had dreamed and found that it seemed to be the correct solution.

**ACTIVITY 9.6** ***A review so easy you could do it in your sleep!***

Can you still remember the answers to these questions?

1) Name the six basic lobes of the cerebrum: ______________________________________________

_____________________________________________________________________________

2) What does the medulla oblongata do? ___________________________________________

3) What does the pons do? _______________________________________

4) Which brain part controls hunger and thirst? __________________

5) The astrocyte and the oligodendrocytes are examples of ___________ cells.

6) The vertebrae in the neck are called the ________________ vertebrae.

7) The vertebrae numbered L1 though L5 are called the ______________ vertebrae.

8) The nerve that connects the eye to the brain is called the __________ nerve.

9) The nerve that connects the ear to the brain is called the ___________________ nerve.

10) The nerves throughout the body are part of the ____________________ nervous system.

11) The brain and spine make up the ______________ nervous system.

12) Is the sympathetic system part of the voluntary or involuntary system? _________________

13) The thin protective membranes surrounding the brain and spine are called the _________________

14) REM stands for _________ _____ ___________________

15) The pineal gland produces this hormone: __________________, which helps you sleep.

16) Which brain waves do you produce while awake? Mostly _____________ but maybe some ____________.

17) When you are in your deepest sleep, your brain is producing which kind of wave? _________

18) How many stages of sleep are there? _____

19) Sleepwalking occurs when your brain is in stage ____.

20) What famous poet wrote a poem while he was asleep? _______________________

# CHAPTER 10

## BRAIN DOCTORS

I take my baby to a pediatrician.
Goo!

There are many kinds of doctors. The doctors you go to see for check-ups and for treatment of common ailments such as sore throats and ear infections are called *general practice* doctors or *family practice* doctors. If you go to a doctor who sees only children as patients, he or she would be a *pediatrician* (the "pedia" part of the word means "child"). If one of these doctors thinks you may have a problem somewhere in your nervous system, they might send you to a doctor who deals only with nervous system problems: a ***neurologist***. If this doctor happens to treat only children, he or she would be a ***pediatric neurologist.*** The neurologist has a general knowledge of the body, just like the family doctor or pediatrician, but he or she stayed in medical school for additional years, studying just the nervous system. Any type of doctor who studies and treats just one kind of problem is called a ***specialist***. A neurologist is a specialist who specializes in the nervous system. Even in the field of neurology, there are specialists. Usually, they still call themselves neurologists, but add "specializing in [whatever]." (Informally they are sometimes called "super specialists.") Specializing in just one thing allows the doctor to take the time to learn everything there is to know about that one topic. For example, a neurologist might specialize in a disease such as epilepsy. This means they could spend all their time learning about just epilepsy, and become very good at diagnosing and treating it. They would see only patients with epilepsy.

Hello. I am a pediatric neurologist. I help children who have problems with their nervous system.

If a nervous system problem requires surgery, the patient will be operated on by a ***neurosurgeon***. This type of surgeon specializes in operating on the brain and spinal cord. Some neurosurgeons specialize even more, and do nothing but brain surgery. (Dr. Ben Carson, mentioned back in chapter 6 was a pediatric neurosurgeon.)

Brain problems that do not affect the function of motor or sensory nerves, and do not require surgery, can often be handled by a doctor called a ***psychiatrist***. A psychiatrist is a medical doctor who specializes in brain disorders that affect mainly behavior, but also might include physical symptoms such as weight gain or loss, fatigue, or rapid pulse. A psychiatrist's office might not look like a typical doctor's office: no thermometers, no latex gloves, no exam table. The psychiatrist talks to his patients and listens carefully as they describe how they are feeling. They will determine if the patient might have a chemical imbalance in their brain that is causing the symptoms. If so, they might prescribe a medication. If not, they might refer the patient to another type of doctor, a ***psychologist***.

A psychologist cannot prescribe medicines. He or she simply talks with the patient and helps them find ways to change their thinking or their lifestyles in order to reduce the symptoms they are experiencing. For example, if the depression a patient is experiencing is not because of neurotransmitters, but is the result of difficult relationships with family members, the psychologist can help the person come up with ways of improving those relationships. Everyone has times in their life when they experience difficulty, and it can be very helpful to talk to someone during those times. Psychologists have been trained to know what to say and to know what advice to give. Sometimes a psychologist will talk to more than one person at a time. The whole family might need a group meeting with the psychologist in order to work out their differences. This is sometimes referred to as ***group therapy***.

One of the first scientists to do research in the field of psychology was ***Sigmund Freud***. Born in Austria, in 1856, Freud's theories about the workings of the human mind came to have a powerful impact on the whole world. He is best known for his theories about the unconscious mind and how it can repress (intentionally forget) bad memories. He believed that dreams are very important and can be used to diagnose mental problems.

He was one of the first doctors to encourage patients to talk openly about their thoughts, feelings, and memories. He might get the conversation started by asking the patient what they dreamed last night, then he would let the patient ramble on about anything they had on their mind. He called this "free association." As Freud listened to them talk, he would be thinking of possible reasons for the patients' thinking patterns. He would often see problems in adulthood as being related to experiences the person had as a child.

Some of Freud's theories were good, and others were pretty crazy. Most of Freud's ideas are not taken seriously anymore, but in the first part of the 19th century, his opinions ruled the world of psychology. Freud became so famous that even today almost all cartoon caricatures of psychologists, are based on Freud.

* * * * * * * * * * * * * * * * * * * * * * * * * * * * * * * * * * * * * * * * * * *

**ACTIVITY 10.1 Ink blots**

Back in the 1910s, a psychologist named Hermann Rorschach (pronounced "roar shock") came up with the theory that people project their own feelings onto other people or things. For instance, if you are angry, you might think that other people look angry, too. Anywhere you would look, you would imagine angry faces. Rorschach designed a series of abstract ink blot designs to show to his patients. He would ask his patients what they thought the blobs looked like. There wasn't any right answer. He believed they would see something relating to the way they were feeling. By listening to their answers, Rorschach believed he could accurately diagnose what was going on in their minds. Ink blot tests were taken very seriously during the early and mid 1900s. They could even be used as evidence in a trial! Nowadays, they are no longer taken seriously (just like Freud) and are seen more as an amusement than a serious science tool.

The actual ink blots that Rorschach used were intended to be permanent secrets because no one was supposed to see them before taking the test. Somehow, though, the very first design in his set was "leaked" to the public and can be seen all over the Internet. We saved you a trip to the Internet and printed it right here. What do you see? An angry elephant? A bat? A butterfly? A jack-o'-lantern face? A mad dog? Two dogs looking in the opposite direction? None of these?

Just for fun, here is an ink blot test we made up. What do you see in each blot? After taking it yourself, you might want to have several other people try it and see how their answers compare to yours. Remember, there is no right answer. Whatever you see is your right answer. Each person's brain has its own interpretation.

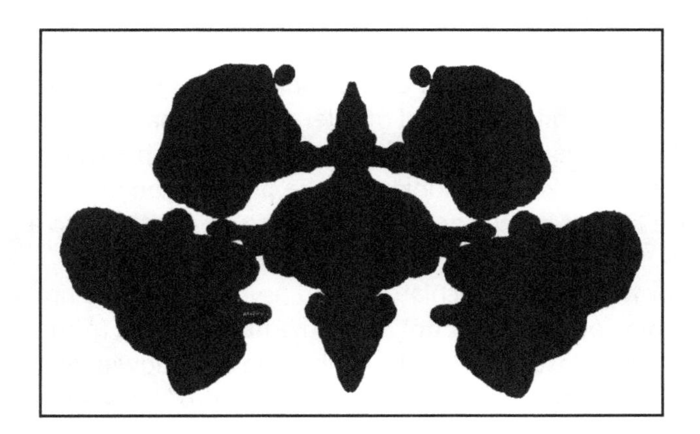

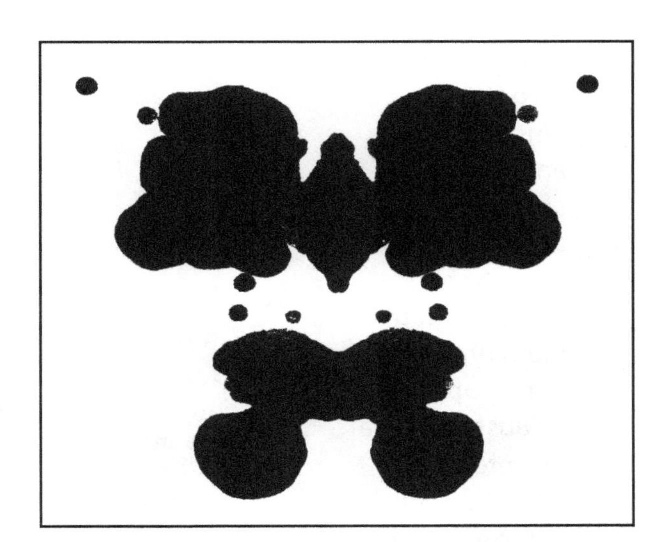

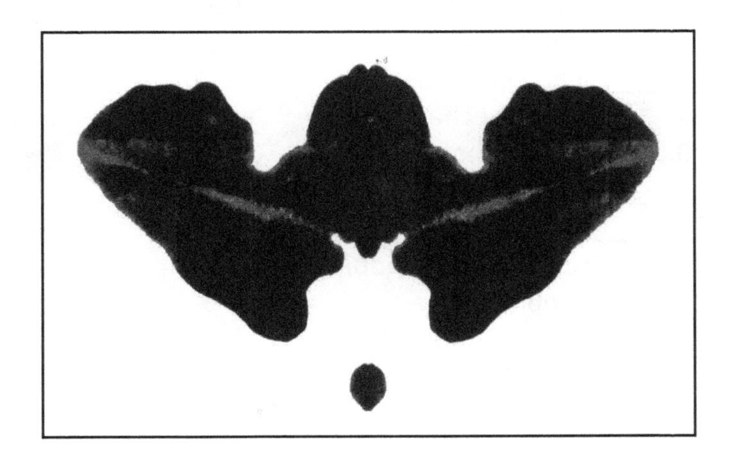

## ACTIVITY 10.2 Phobias

We all have fears. Some fears are good and they keep us safe. Being afraid of lightning and thunder keeps us from walking out into a storm and getting hit by a lightning bolt. But if our fears get out of control and become so strong that they are no longer helpful, we call them ***phobias***. For example, someone who has a phobia of spiders doesn't just avoid them while gardening—they might refuse to go into a house just because they saw a tiny spider crawl past the front door two weeks ago! A phobia of large animals might make it impossible for you to visit a zoo or take a walk in the woods.

If you want to get rid of a phobia, you might go to a psychologist for help. One technique the psychologist might use is called *desensitization,* a process by which you gradually build up a level of tolerance for what you are afraid of. If you are afraid of dogs, you might start by just saying the word "dog." Once you feel comfortable with that, you might move up to looking a picture of a dog. When you are okay with looking at the picture, you might look at a real dog, but from far away. Bit by bit, you would take little steps closer to being able to be right next to a dog without being afraid.

Scientists love to come up with complicated names for things, and psychiatrists and psychologists are no exception. There is a fancy name for almost any fear you can think of. Most, but not all, of the names come from Greek or Latin. See if you can match the phobia name on the left its description on the right. Use any knowledge you have of other words that look or sound similar. (For example, where have you seen the word "dendro" before? Do you know what a "feline" is? What other words do you know that have "hydro" in them?) Good luck!

1. _____ achluophobia
2. _____ acrophobia
3. _____ amnesiophobia
4. _____ androphobia
5. _____ apiphobia
6. _____ arachnophobia
7. _____ autophobia
8. _____ bacteriophobia
9. _____ bibliophobia
10. _____ brontophobia
11. _____ claustrophobia
12. _____ cleptophobia
13. _____ dendrophobia
14. _____ entomophobia
15. _____ felinophobia
16. _____ hemophobia
17. _____ hydrophobia
18. _____ lilapsophobia
19. _____ logizomechanophobia
20. _____ olfactophobia
21. _____ polyphobia
22. _____ pyrophobia
23. _____ sesquipedalophobia
24. _____ toxiphobia
25. _____ triskaidekaphobia
26. _____ xenophobia

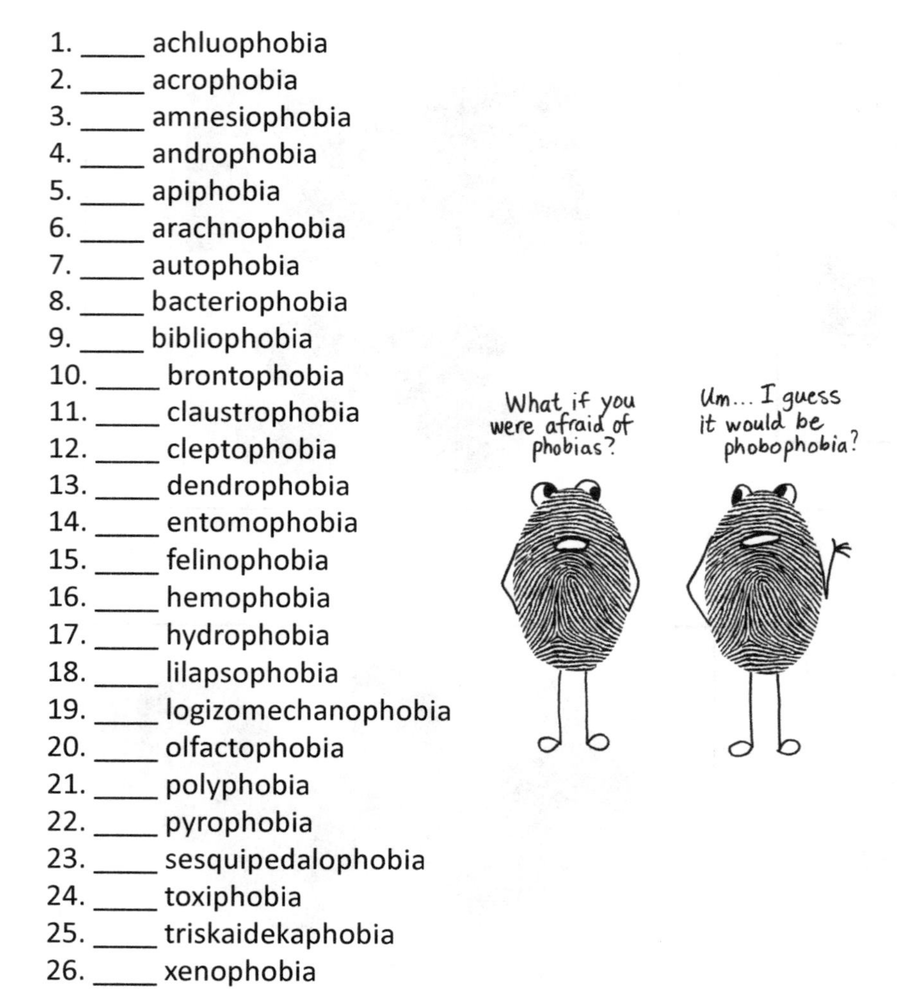


A. fear of being alone
B. fear of poison
C. fear of the number 13
D. fear of computers
E. fear of heights
F. fear of blood
G. fear of strangers
H. fear of more than one thing
I. fear of fire
J. fear of men
K. fear of the dark
L. fear of cats
M. fear of bees
N. fear of trees
O. fear of small, closed-in spaces
P. fear of bad smells
Q. fear of water
R. fear of things being stolen
S. fear of getting amnesia
T. fear of spiders
U. fear of thunder and lightning
V. fear of insects
W. fear of hurricanes and tornadoes
X. fear of germs
Y. fear of books
Z. fear of long words

# CHAPTER 10.5

## BRAIN PROBLEMS

**NOTE: This chapter is optional. If reading about diseases is distressing, skip it.**
**You can always come back to it later if you need the information.**
**However, there is nothing gross, graphic, or inappropriate—have no fear!**

Like any part of the body, the brain can get sick of sustain various types of injuries. Now that you know a bit about how the brain works, you might be able to predict some of these problems. Any part of the system could malfunction: the neurons, the glial cells, the myelin insulation on the axons, the blood-brain barrier, the blood supply to the cells, the synapse gaps between the cells, etc. Let's take a look at a few common problems and see what doctors can do to fix them.

The most common problem associated with the head is a ***headache***. The brain, however, has no pain sensors at all—it *can't* hurt. So what hurts when you have a headache? It sure feels like your brain is hurting sometimes. Since it can't be the brain, it has to be the tissues surrounding the brain. There is a massive network of blood vessels around and below the brain, and if these vessels suddenly enlarge this can result in a sensation of pain. (Some migraines fall into this category, but some don't.) There are medicines that can shrink the blood vessels a bit so the throbbing sensation goes away. Even caffeine will sometimes achieve this, so some people will drink a cup of coffee to cure their headache. ***Migraines*** are headaches that usually occur on one side of the head and often come with other symptoms, too, like dizziness, feeling nauseated, or sensitivity to light. Migraines can have any number of causes. Usually there are specific "triggers" that get them started. Common triggers include foods, stress, exercise, bright lights, or changes in the weather. Migraines, as well as other types of headaches, can also be caused by injury to, or misalignment of, the cervical vertebrae.

Other types of headaches hurt because of tightness in the muscles of the neck or eyes. Medicines like aspirin, acetaminophen or ibuprofen stop the sensation of pain by interfering with the process of pain signals being sent to the brain. If the brain never gets the pain signals, the result is that you don't "feel" the pain. Headaches can also be caused by food sensitivities and allergies, sudden drops in blood sugar, and, rarely, brain tumors. People who don't know the causes of their headaches can be greatly helped by ***functional medicine*** or ***integrative medicine*** doctors. This type of doctor likes to play detective and figure out what is going on. They take everything into consideration, including your diet, your environment, your genetics, your activity level, and even your relationships. They will prescribe medicines if necessary, but will first try to figure out what is causing your headaches and if there is a way to stop them by making changes to lifestyle or diet.

If you smack your head against something, you might get a ***concussion***, which is sort of like a bruise in the brain. Microscopic blood vessels (capillaries) can break, allowing blood to seep out. For a mild concussion, there isn't much you can do. You just rest a bit, then make sure it doesn't happen again. Those glial cells will get to work and clean up the mess. For a severe head injury, doctors may need to drain off extra blood and fluids so they don't cause too much pressure inside the skull. You can do a lot to prevent severe head injuries by wearing safety devices such as helmets and seat belts.

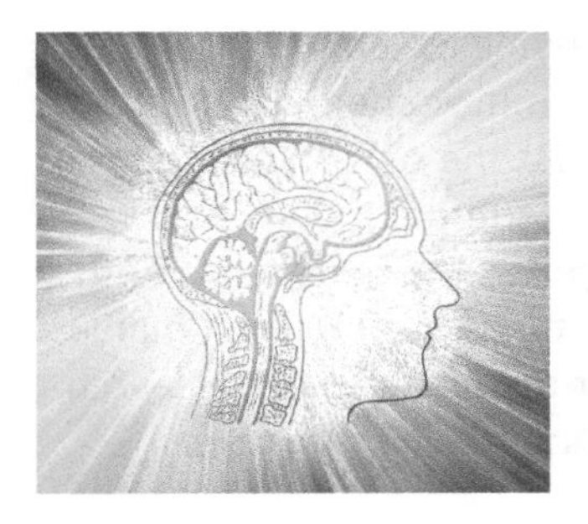

If the neurons in the brain make too much electricity, an "electrical storm," will result. In the disease called ***epilepsy***, one part of the brain suddenly starts producing way too much electricity, causing a ***seizure*** (see-zhure). There are different types of seizures, depending on what part of the brain is sending out the signals. A seizure can be hardly noticeable—the person just sits and stares blankly for a few seconds— or it can be very severe, causing the person's whole body to twitch and jerk. A seizure can be as short as few seconds, or as long as an hour. Most last a few minutes. Seizures can happen as often as several times a minute or as rarely as a few times a year.

If a neurologist suspects that a person might have epilepsy, they order an ECG test. (This is the same machine we learned about in the chapter on sleep.) Electrodes are taped to the head at various places and the patterns of electricity put out by the brain are recorded as squiggly lines on a graph. If the person has a seizure, it will show up as spiky lines on the graph, going both higher and lower than the regular squiggly lines.

The most common treatment for epilepsy is medicine that stops the brain from making so much electricity. These drugs are called ***anti-convulsants.*** (To "convulse" means to have a seizure). Most people with epilepsy find that these medicines work very well, allowing them to live normal lives. In very rare cases, surgery can help. This was the case with the "split brain" patients we learned about in chapter 3.5.

The most common brain problem in older people is called a ***stroke***. We mentioned this brain injury way back in chapter 1.5, when we learned that brain injuries were a key to learning the function of brain parts. When someone has a stroke, one of the blood vessels in their brain becomes blocked by a blood clot, preventing blood from reaching that area of the brain. Without blood, brain cells die. The symptoms of stroke depend on what area of the brain is affected. If blood is cut off to the motor cortex, one side of the body could be paralyzed. If the blockage is in the temporal lobe, near the speech center, the person may be unable to talk. Strokes can range from so mild that it is barely noticeable, to so severe that it causes death. Fortunately, the brain has great powers of recuperation, and can re-route neuronal networks around damaged areas. Given time, and lots of strengthening exercises, stroke patients can make substantial recovery. Doctors give stroke patients medicines that break up blood clots and prevent more from forming. (Your chances of having a stroke can be greatly reduced if you eat a low-sugar diet, exercise regularly, and don't smoke.)

Neurons are the only cells in the brain that do not reproduce. The other cells are constantly making copies of themselves so that when the original cells die there are new ones to take their place. Usually, cells reproduce at a slow, steady rate. Very rarely, one of the cells will get out of control and start reproducing so fast that a huge clump is formed. These fasting-growing cells don't know when to stop, and the clump gets bigger and bigger. The scientific name for this fast-growing clump of cells is a ***tumor***. When a tumor occurs in the brain, it is most often one of the glial cells that has gone bad. Tumors originating from neurons are not impossible but are very rare. Tumors are much more likely to start in any of the glial cells, or in the cells of the meninges or the blood vessels.

Some tumors will stay to themselves and not infect any other body parts. These tumors are called ***benign*** *(bee-NINE)*. Other tumors will send some of their cells to other parts of the body where they will teach those body parts to begin doing the same thing—growing and reproducing too fast. This type of tumor is called ***malignant*** *(mal-LIG-nant)*. In either case, the doctor may try to remove the tumor with surgery. In addition, the doctor may use medicine or radiation to kill off any remaining out-of-control cells. MRI scans are very helpful to doctors when they are diagnosing and treating tumors of the brain.

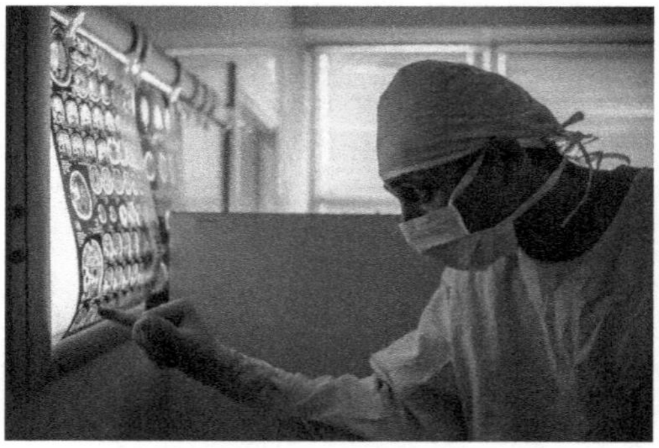

Some brain problems fall into the category of ***mental illness***. These are physical problems that are often made worse by situational or environmental factors. The symptoms often affect behavior, especially social behavior, but can include non-brain symptoms, also. One of the most common problems is ***depression***. Someone with depression doesn't just feel sad, their whole body feels bad. They may feel so tired that they can't get out of bed. They could have body aches or headaches, and they might stop eating (or eat to much). The treatments most often used are medicines that increase the amount of neurotransmitter chemicals in the synapses. The opposite problem is called ***mania***. The signals in the brain flow so fast that the person has too much energy and too many thoughts. They may talk non-stop, hardly sleep, and feel so energetic that they think they can do almost anything, even very unrealistic things. In a condition called ***bi-polar***, the person has both depression and mania, alternating back and forth. Medicines used for epilepsy are sometimes used to treat mania and bipolar disorder. They stabilize the brain and keep the activity from going too low or too high. Recent research has suggested that both depression and mania could be related to the activity of mitochondria in brain cells.

***Schizophrenia*** *(SKITS-oh-FREEN-ee-ah)* is a mental illnesses characterized by the inability to tell the difference between reality and imagination. Schizophrenics often see and hear things that aren't real (hallucinations) or think they have magical powers such as mind reading or talking to spiritual beings. They usually have trouble with sleeping, organizing, and concentrating, as well. No one knows for sure what causes this illness, but genetics seem to be part of it. It is treated with medicines called "anti-psychotics" *(sie-KOT-iks)*. Some of these medicines aim to reduce the amount of a neurotransmitter called dopamine.

The most famous person who suffered from schizophrenia was the artist Vincent Van Gogh. (His painting "Starry Night" is shown here.)

***Cerebral palsy*** is something you have from birth. It's not a mental illness. It is caused by damage to the cerebrum before, during, or right after birth. The word "palsy" means that the muscles don't function properly as a result of the brain damage. A person with cerebral palsy may walk awkwardly or have trouble speaking clearly. In severe cases, the person may be confined to a wheelchair, or be unable to dress or feed themselves. Sometimes they can experience seizures that cause muscles groups to contract repeatedly. Rarely, the palsy affects the frontal lobe and causes mental retardation, but in most cases the frontal lobe is completely normal. A person with cerebral palsy can be extremely intelligent, despite their difficulty walking and talking. Many people with cerebral palsy have jobs and careers and families, just like everyone else.

Another problem that can occur at birth, or shortly after, is ***hydrocephalus*** *(HI-dro-SEF-el-us)*. "Hydro" means "water" and "cephalus" means "head," so these babies have too much water (fluid) inside the head. You'll remember than the ventricles (empty spaces) deep inside the head are filled with cerebrospinal fluid. The ventricles are supposed to drain down through the spine, and out the bottom. If the "drain pipe" becomes blocked, the cells in the ventricles go right on producing fluid, and all that extra fluid has no place to go. The ventricles start to swell because of the extra fluid. The enlarged ventricles push on the brain and make the whole head expand. Needless to say, this is not good. Doctors must clear the blockage and get the ventricles to drain properly. Sometimes they put in an artificial drain pipe called a ***shunt***. They can leave this drain pipe in for a long time, even years, to make sure the ventricles don't get clogged again.

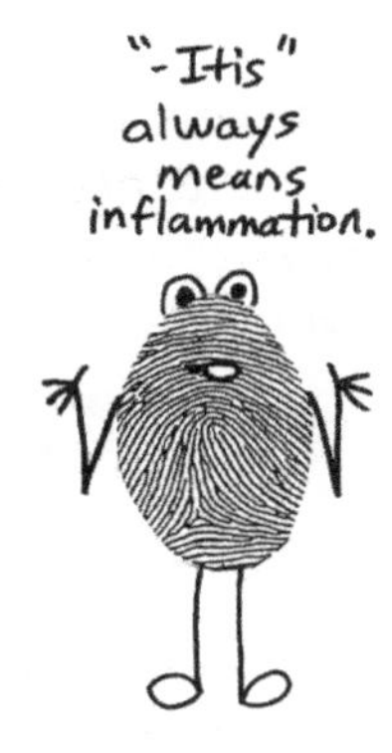
"-Itis" always means inflammation.

If something goes wrong with the blood-brain-barrier, and a bacteria or virus does get into the brain, it can cause a condition known as ***encephalitis*** *(en-SEFF-el-ITE-is)*. When you see "-itis" on the end of a word it means "inflammation." Encephalitis is inflammation (swelling) due to an infection inside the head. This causes fever and headache, usually accompanied by mental confusion. For example, the person might try to pick up objects that are not really there, or they may say things that don't make sense. Hospitals can treat encephalitis with special antibiotics that can get through the blood-brain-barrier. Fortunately, encephalitis is very rare, and certainly isn't something you need to worry about next time you get a fever with a headache. Many normal viruses also cause fever and headache, and the odds are good that you'll have one of these, not encephalitis.

Some brain problems occur as a result of getting older. Age-related diseases that cause problems with thinking and remembering are known as ***dementia*** *(de-MEN-shah)*. One type of dementia that is in the news a lot is ***Alzheimer's disease***. In Alzheimer's disease, neurons begin dying. Since you can't replace them, this means that those neurons are gone for good. The empty space where they used to be is filled in with cerebrospinal fluid. You wouldn't notice if even a thousand neurons died, but if millions start dying, you lose neuronal networks that you spent a lifetime building up. Memories and knowledge disappear. People with Alzheimer's generally get worse as time goes on, but there are new medicines being developed that can slow down the progress of the disease. Researchers are also looking into ways that the disease can be prevented, possibly with diet and exercise, or by taking small amounts of medicine that protects the neurons from damage. Alzheimer's research is becoming a very important branch of medical science and a lot of progress has been made in understanding how the disease works and what can be done about it.

Sometimes, age-related diseases turn out to have an infectious component to them. Researchers studying Lyme Disease examined the brains of people who (supposedly) died of Alzheimer's, and they discovered the bacteria that causes Lyme in many of the brains. Was this a coincidence? More research is needed.

A very common disorder that you may already be familiar with is ***Autistic Spectrum Disorder.*** This term, ASD, is a fairly new category that attempts to cover a broad range of related conditions, some mild and some severe. The word root "auto" means "self," so autistic disorders might have some overlap with narcissistic personality disorder, but to a psychiatrist or psychologist, they are distinctly different disorders. People with ASD lack the ability to see the world through someone else's eyes, but they are generally not manipulative. In fact, they are often quite innocent and they rarely commit crimes. People with any type of autism have trouble "reading" faces, so they often miss social cues that their normal friends pick up. For example, they can't tell if the person they are talking to is really interested in what they are saying. They have to be taught what to look for in facial expressions and body language.

In severe cases, autistic children can't even talk. They don't laugh or smile like other children, and spend much of their time doing repetitive actions that they find soothing. Most people with severe cases get better over time (with therapy), but some remain very disabled. Even people with disabling autism, however, sometimes have amazing abilities in one specific area (like the savant syndrome from chapter 3).

A mild form of autism is called ***Asperger's Syndrome***. The mildest cases often go undiagnosed and although society may see these people as a bit strange, they somehow find a niche for themselves and manage to live a fairly normal life, especially if they happen to have talents that people value.

The most common brain problems are on the very edges of the mental illness category, but tend to be treated as a separate category: ADD and ADHD.

**ADD** stands for **A**ttention **D**eficit **D**isorder. People with this disorder have a hard time concentrating or staying "on task." A student with ADD might not be able to pay attention in class. They might daydream when they are supposed to be completing an assignment. They are often forgetful and are constantly losing important items such as their keys, their phone, their wallet, their coat, etc. Sometimes treatment involves using medications, but many people simply decide to find ways to alter their behavior in order to minimize the impact of ADD.

**ADHD** stands for **A**ttention **D**eficit **H**yperactivity **D**isorder. People with this disorder have some of the same issues as people with ADD, but in addition they also have a jittery brain that takes their thoughts off in too many directions, making it hard not only to concentrate, but often even to sit still. Children with this disorder can be very disruptive in a classroom. Even with normal intelligence, they can have trouble learning. ADHD is believed to occur in the prefontal cortex area, due to a reduced supply of the neurotransmitters dopamine and norepinephrine. Treatment usually involves taking medications, one of the most common being Ritalin, which prevents the recycling of these neurotransmitters so that there is a higher concentration of them in the synapses.

***Personality disorders*** also lie on the edge of the category of mental illness. Some are severe enough that they do merit being called a mental illness. Others are mild, and the person might go through their whole life without ever being diagnosed. Personality disorders are thought to be caused by a combination of genetics and environment. You are born with various tendencies, both strengths and weaknesses. Life circumstances will then shape you, causing certain traits to become more or less dominant. Sadly, abuse during childhood is all too often a major factor that contributes to developing a personality disorder. Personality disorders are the most difficult mental problems for doctors (and for society) to deal with because the person usually either does not know anything is wrong, or is unwilling to accept treatment. People who commit randoms acts of violence on total strangers (school shootings, etc.) almost certainly suffer from one of these syndromes.

***"Cluster A"*** personality disorders include ***paranoia*** and ***schizoid*** disorder. People with paranoia develop irrational fears that everyone is "out to get them." They distrust everyone. In the worst cases, they become hostile to people because of these unfounded and irrational fears. Schizoid people have some of the same traits as schizophrenics, but develop less severe symptoms. They lack social skills and tend to be loners. They often appear "cold," having little or no emotion. They usually dress oddly and may say inappropriate things during conversations.

***"Cluster B"*** personality disorders include ***antisocial, borderline, and narcissistic*** (nar-si-SISS-tick). People with **antisocial personality disorder** are aggressive, defiant, argumentative, angry, and compulsive (no self restraint). They are likely to break laws and commit crimes. When caught, they show no remorse for their actions.

**Borderline personality disorder** is harder to define. The word "borderline" refers to the fact that this The main characteristic of this disorder is instability. Borderlines often have sudden mood swings. For example, within minutes they can become so enraged that they say and do things that damage relationships. They are often very fearful, as well, and can be afraid of being alone. They might say or do anything to manipulate someone into not leaving. They are often impulsive and engage in risky behaviors. They often have low self-esteem and are prone to developing addictions. This disorder used to be seen as untreatable, but that is not true anymore. Many people have made great improvements through treatments offer by both psychiatrists and psychologists.

***Narcissistic personality disorder*** is selfishness to the extreme. (Narcissus was a mythological Greek figure who fell in love with his reflection in a pool and spent the rest of his life staring at it.) We all struggle with a normal amount of selfishness. As babies and toddlers we thought the world revolved around us. We had to be taught how to share our toys. We may have been taught the Golden Rule, "Treat others as you, yourself, want to be treated." A normal brain has the capacity to learn these things. There appears to be a part of our brain that allows us to imagine how things look from another person's perspective. We call this "empathy." People with NPD can't do this; they seem to have no ability to empathize. Even as adults, they are unable to put the needs of others ahead of their own needs. If they seem to show humility, it is just an act in order to manipulate others in to doing their bidding. Rarely, people with NPD can overcome their disorder, but most often they go untreated because they think it's everyone else who has a problem, not them.

Some people with antisocial personality disorders can be classified as being either a "**sociopath**" *(SO-see-oh-path)* or a "**psychopath**"*(SIE-ko-path)*. Both disorders involve the person having no ability to understand how another person is feeling. PET and MRI scans of their brains reveal areas that don't look normal. As children, psychopaths are able to harm animals or younger children with no remorse. As adults, they learn to manipulate everyone around them in order to provide benefit to themselves. They don't care what the consequences are for other people. Sadly, some turn into killers whose crimes end up on the front page of newspapers. Sociopaths are the "light" version of the disorder and are less likely to commit crimes. Scientists estimate that 1% of the population has some type of narcissistic or sociopathic disorder.

**"Cluster C"** disorders involve intense anxiety and fear, and include "avoidant" disorder (extreme avoidance of things that are fearful), "dependent" disorder (extreme dependency when there doesn't seem to be any rational explanation for the dependency), and "obsessive-compulsive disorder" (OCD).

People with ***OCD*** have brains that concentrate too much, not too little. The brain concentrates so much on a few things that all other thoughts get pushed aside. This type of hyper-concentration is called "obsession." Commonly, people with OCD obsess about neatness, cleanliness or daily routines. For example, one woman straightened the mugs in her cupboard so all the handles pointed the same direction, but an hour later she was doing it again, then again a few hours after that. Her brain was in a continuous loop, never being able to accept the fact that this chore had been completed. People with OCD are often afraid of germs and will wash their hands so many times each day that their skin will be red and painful. Treatment for OCD usually combines medications with behavioral therapy.

## ACTIVITY 10.3 Make the diagnosis

The doctor's in... and it's you! For each of the following patients, give your opinion of what might be wrong with them, based on what you read in this chapter.

**PATIENT #1:** Age 72. Often gets in the car, then forgets where he was going. Has been seeing staring blankly at items on his desk, not knowing what to do with them. MRI scans show large ventricles, which could be places where dead brain cells used to be.

A possible diagnosis: ___________________________

**PATIENT #2:** Age 1 year. This baby had a difficult birth. She is having trouble learning to walk and seems unable to control her legs. She seems to have normal intelligence, just trouble with motor skills.

A possible diagnosis: __________________________

**PATIENT #3:** Age 10. Several times a day the patient stops what she is doing, moves around in a funny way for a few seconds, then seems fine again. An EEG machine recorded some unusual electrical activity.

A possible diagnosis: ________________________

**PATIENT #4:** Age 65. Experienced sudden loss of control over one arm. One minute he was fine, the next minute he could not move his arm. An EEG machine recorded normal brain activity.

A possible diagnosis: _______________________

**PATIENT #5:** Age 34. Complains of fatigue, loss of appetite, and feelings of despair. All blood tests have come back negative (meaning normal).

A possible diagnosis: __________________________

**PATIENT #6:** Age 6 months. Baby's head is too large on one side. MRI scans shows no tumors.

A possible diagnosis: _____________________________

**PATIENT #7:** Age 24. Frequent headaches, but always on Friday or Saturday nights. Fine during the week. If asked to stretch her neck, the muscles seem very tight.

A possible diagnosis: ____________________________

**PATIENT #8:** Age 12. The patient has had a mild headache for three days and is having trouble concentrating in school. The patient's mother says he was hit in the back of the head with a baseball three days ago.

A possible diagnosis: ____________________________

## ACTIVITY 10.4 Brain surgeries where the patient is awake—and playing music!

In some brain surgeries, the patient needs to be awake because the surgery is in a delicate area next to some motor nerves and the surgeon wants to make sure those motor nerves are not being damaged. (Remember, the brain itself doesn't feel any pain.) Depending on which part of the brain is being operated on, the surgeon may ask the patient to answer questions, move their arms or toes, etc. In rare cases, the patient is a professional musician and the surgeon wants to make sure their ability to play their instrument remains intact after the surgery. Because it looks so funny to see someone playing the banjo, guitar or violin while their brain is being operated on, these operations are often filmed, and some of these videos get posted on YouTube or other video streaming services. Start with these key words: "playing banjo during brain surgery." This will probably return a video (of famous banjo player Eddie Adcock) that doesn't actually show the brain, so there is nothing gross. If you like that one, you could try adding "guitar" or "violin" to key words "during brain surgery."

# BIBLIOGRAPHY

These are the main books I used for my research (2007, 2014), listed in order of how much I used them:

The Brain and Central Nervous System, from the Reader's Digest series: *Your Body, Your Health*. Published in 2002 by Reader's Digest in Pleasantville, NY. ISBN 0-7621-0436-8

Big Head! by Dr. Pete Rowan. Published in 1998 by Alfred A. Knoff, NY. ISBN 0-679-89018-1

The Human Nervous System; An Anatomic Viewpoint, by Murray L. Barr. Published in 1979 by Harper & Row, Publishers. ISBN 0-06-140312-1

Inside the Brain, by Ronald Kotulak. Published in 1996 by Andrews McMeel Publishing, Kansas City. ISBN 0-8362-3289-5

How the Brain Learns, by David A. Sousa. Published in 2006 by Corwin Press, Thousand Oaks, California. ISBN 1-4129-3661-6

Mapping the Mind, by Rita Carter. Published in 1998 by University of California Press, Berkeley and Los Angeles. ISBN 0-520-22461-2

The Art of Changing the Brain, by James E. Zull. Published in 2002 by Stylus Publishing, Virginia. ISBN 1-57922-054-1

Right-Brained Children in a Left-Brained World, by Jeffrey Freed and Laurie Parsons. Published in 1997 by Fireside Books, a division of Simon & Schuster, NY. ISBN 0-684-84793-0

These are the main websites I used for the 2014 edition.
Also not listed are some YouTube videos I watched, although some of them are posted on the playlist.

**Wisconsin Medical Society at www.wisconsinmedicalsociety.org**

**Howstuffworks.com**

**"Neuroscience for kids" at www.facutly.washington.edu/chudler/neurok.html**

**Wikipedia, the free online encyclopedia at: www. wikipedia.org**

For the 2022 update, I checked facts using many Wikipedia articles, as well as using information from some of the videos that are posted on the Brain Curriculum playlist on YouTube.com/TheBasementWorkshop.

# ANSWER KEY

# ANSWER KEY

## CHAPTER 1

### ACTIVITY 1.1

1) CT 2) MRI 3) PET 4) MRI is dangerous for anyone with metal implants (because of the strong magnetism used) 5) PET, because it is the only one that can show activity level in real time.

## CHAPTER 1.5

### ACTIVITY 1.5

MRI:
1) Magnetic Resonance Imaging
2) Very large! The interior hole of an MRI is large enough for a person's entire body to go into it.
3) Yes, it has an incredibly powerful magnet inside.
4) Answers will vary. MRI is used to diagnose any problem that involves healthy versus unhealthy body tissue, such as tumors, joint deterioration, blockages, torn ligaments, heart disease, multiple sclerosis, and others. MRI can also be used in non-medical application such as determining the physical properties of rocks and other non-living materials.
5) Answers will vary. The strong magnet is used to line up protons in hydrogen atoms. Then, a radio frequency is pulsated in the area that is to be examined. The radio waves will cause some of the hydrogen atoms to absorb energy (causing "resonance"). When the radio waves are stopped, this extra energy is released, and is detected by the imaging system, which displays it on a screen.
6) fMRI can give 3D images and is used to detect changes in oxygen level in the tissues. fMRI studies metabolism and is more like PET in this regard. MRI only shows you structure, not functioning.
PET:
1) Positron Emission Tomography
2) Patients must receive an injection of radioactive glucose (sugar).
3) The PET scan is used to show the level of activity in bodily tissues (as opposed to MRI which shows structure, not activity). In other words, PET scans show how quickly cells are using glucose. Normal tissue and diseased tissue use glucose at different rates and this shows up on PET scans as different colors.
4) The PET scan detects radiation given off by the radioactive glucose. The more radiation, the brighter the color. Cells that are more active will have more of the radioactive glucose in them, therefore they will appear as reds and yellows. Less active areas will be blues and greens.
5) PET is used to diagnose many heart conditions and blood flow problems. It can also be used to search for tumors, since they take up glucose at a faster rate than normal tissue.

## CHAPTER 2

### ACTIVITY 2.1

ACROSS: 3) corpus callosum, 5) capillaries, 7) frontal, 8) white, 9) sensory, 10) parietal, 11) stem, 14) temporal, 15) cortex, 16) blood
DOWN: 1) occipital, 2) cerebellum. 4) cerebrospinal fluid, 6) cerebrum, 12) motor, 13) gray
BONUS QUESTION: left

### ACTIVITY 2.4

2) Playing the piano
FRONTAL LOBE: Decides to play, understands what you are playing, send signals to motor cortex to move fingers, interprets what the musical notation means ("reads" the music)
MOTOR CORTEX: Sends signals to fingers (and to feet if you are using pedals)
SENSORY CORTEX: Feels the keys under your fingers
PARIETAL LOBE: Senses where your hands are arms are in relationship to the piano, senses that you are sitting in front of the piano, helps keep you balanced on your piano bench
OCCIPITAL LOBE: Sees the notes on the sheet music
TEMPORAL LOBE: Hears the notes as they are being played

3) Riding a bicycle
FRONTAL LOBE: Decides to ride, knows where you are going and makes decisions about whether to turn left or right, sends signals to motor cortex,
MOTOR CORTEX: Sends signals to muscles in arms and legs
SENSORY CORTEX: Feels the wind on your skin, feels the pedals under your feet, feels the hand grips in your hands
PARIETAL LOBE: Helps you keep your balance on the bike
OCCIPITAL LOBE: Sees the landscape around you
TEMPORAL LOBE: Hears things like birds singing, traffic sounds, the wind in your ears

4) Talking on the phone
FRONTAL LOBE: Decides what to say and when to say it, sends signals to the motor cortex
MOTOR CORTEX: Sends signals to the muscles in the mouth and vocal chords (and your hand that is holding the phone)
SENSORY CORTEX: Feels the phone in your hand
PARIETAL LOBE: Isn't essential for talking on the phone, but it is doing its basic job of sensing where your body is in relationship to the objects around you
OCCIPITAL LOBE: Isn't essential for talking on the phone, but if your eyes are open, it is taking in the sights around you
TEMPORAL LOBE: Hears the words that are being said to you by the other person, makes sense of the words and sentences you are hearing.

## CHAPTER 2.5

### ACTIVITY 2.4

We measured the cortex on the left to be about 26 centimeters and the one on the right to be about 22 centimeters, but your measurements might be a little more or less. The main point is to notice how folding and creasing allows something to occupy a smaller area. If you straightened out those cortexes, they would extend well beyond the edges of the page!

### ACTIVITY 2.8

1) DURA 2) PIA 3) ARACHNOID 4) FISSURE 5) CORPUS 6) CALLOSUM
7) ANTI 8) CEREBRUM 9) CEREBELLUM 10) GYRUS 11) SULCUS
12) OCCIPITAL 13) CORTEX 14) MENINGES 15) PARIETAL

## CHAPTER 3

### ACTIVITY 3.3

Pennsylvania, Minneapolis, Arkansas
San Diego, California, Tennessee

### ACTIVITY 3.6

The one turned the wrong way is C.

## CHAPTER 4

### ACTIVITY 4.2

ACROSS: 3) motor cortex 7) fornix 8) parietal 11) occipital 13) hypothalamus 14) pituitary 16) midbrain 17) temporal 18) sensory cortex
DOWN: 1) cingulate gyrus 2) amygdala 4) cerebellum 5) medulla oblongata 6) olfactory bulb 7) frontal 9) hippocampus 10) pons 12) ventricles 15) thalamus

## CHAPTER 5.5

### ACTIVITY 5.6

1) fissure 2) gyrus 3) sulcus 4) ventricles 5) homunculus 6) dendrites 7) Schwann cell 8) node of Ranvier 9) nucleus 10) glial cell 11) pons 12) medulla oblongata 13) cerebellum 14) occipital lobe 15) parietal lobe 16) sensory cortex 17) motor cortex 18) frontal lobe 19) olfactory bulb 20) cingulate gyrus 21) corpus callosum 22) midbrain 23) thalamus 24) skull 25) dura mater 26) arachnoid layer 27) pia mater 28) cortex (gray matter) 29) PET scan 30) MRI scan 31) synapse 32) neurotransmitters 33) receptor site 34) vesicles 35) axon knob

**ACTIVITY 5.6**

NOTE: If you came up with a different answer but it makes sense, you can count that, too.
1) "limbic" because the other three are lobes of the cerebrum and the limbic system is not.
2) "synapse" because the other three are words that describe shapes found on the cerebrum and a synapse is a microscopic gap, not a shape.
3) "CRV" is not a type of brain scan.
4) "neuron" because it is not a glial cell.
5) "carbon dioxide" because it is a waste product and the other three are things the brain needs.
6) "Gage" because the other three are scientists. Gage was not a scientist.
7) "medulla oblongata" because it does not play a role in memory.
8) "frontal lobe" because all the others are connected to the senses in some way. The frontal lobe is not directly connected to any senses.
9) "pituitary gland" because there is just one of them, and the other structures all have two sides or two lobes
10) "synapse" because it is not actually part of a neuron. It is nothing but empty space!
11) "homeostasis" because it is not a word associated with the synapse
12) "potassium" because it is not something that directly affects neurotransmitters

**ACTIVITY 5.7**

1) False, it is in the temporal lobe. 2) True 3) True 4) False, Schwann cells are found only in the body, not in the brain. Oligodendrocytes perform the insulation function in the brain. 5) True 6) False, a shunt carries cerebrospinal fluid away, not blood. 7) True 8) False, this is why alcohol can make you intoxicated. 9) True, because more wrinkles means more surface area, which means more cortex which means more area for brain functions. 10) True 11) True, because the BBB is created by tight junctions between cells. 12) False, the neurotransmitters are made inside the neuron. 13) True 14) False, there are four ventricles. Two are matching left and right, two are 15) True

## CHAPTER 6

**ACTIVITY 6.2**

1- 6) frontal, motor cortex, sensory cortex, parietal, occipital, temporal
7) beating of heart, keeping lungs breathing
8) Helps with balance, coordinates movements 9) sea horse 10) the basal ganglia
11) pons 12) occipital 13) temporal lobe 14) motor cortex 15) frontal
16) temporal 17) hypothalamus 18) nose 19) grow 20) dendrites
21) L 22) R 23) R 24) R 25) L

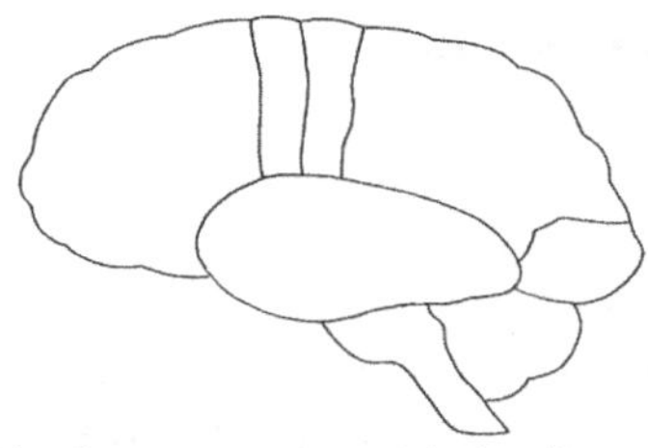

Check your answers with the diagram on page 23

## CHAPTER 6.5

**ACTIVITY 6.4**

**First joke:**
Node of Ranvier, oligodendrocytes, Schwann, neurotransmitters, knobs, myelin, receptor sites
PUNCH LINE: "Hey, kid, you've got a lot of potential!"

**Second joke:**
plasticity, billion, serotonin, network, hippocampus PUNCH LINE: the Neural Network

## CHAPTER 7.5

**ACTIVITY 7.6**

IMPLICIT: how to tie my shoe, how to do a certain dance, how to type on the keyboard without looking, how to zip a zipper, how to sign my name
EPISODIC: the phone number of Carlo's Pizza (could possibly be semantic, if you have known it a long time), my math lesson from last week, how my yard looks in the winter (if it is winter right now and you are remember seeing it today or yesterday), where the Hawaiian Islands are located (if you just learned this in the past few days or so), how my bedroom looks right now, don't mix peppermint with cola, I must be home by 9. my homework assignment for tomorrow, the dog next door bites (assuming you learned the hard way), what I got for my birthday
SEMANTIC: the feel of my dog's fur, 3x3=9, how my yard looks in the winter (if it is a general memory, not a recent memory from today or yesterday), where the Hawaiian Islands are located (assuming you have known it for a while), how to spell the word "brain," the nursery rhyme "Humpty Dumpty," the fact that birds can fly

## ACTIVITY 7.8

1) CHUNKING 2) IMPLICIT 3) PROCEDURAL 4) EPISODIC 5) SEMANTIC
6) EXPLICIT 7) SEVEN

Answer to question: Mental exercise and nutrition

# CHAPTER 8

## ACTIVITY 8.1

| C | F | M | Y | H | L | T | H | B | L | Y | M |
|---|---|---|---|---|---|---|---|---|---|---|---|
| E | Z | E | E | N | S | W | C | L | R | A | R |
| N | D | O | R | L | N | V | T | I | R | W | O |
| B | X | G | A | G | P | S | A | R | G | N | P |
| I | L | S | H | J | L | S | R | M | F | T | F |
| J | T | I | M | O | V | I | C | O | U | G | H |
| O | R | W | N | C | U | D | S | A | L | R | H |
| M | F | S | L | K | H | O | Z | N | G | D | L |
| E | S | P | A | T | E | L | L | A | R | E | Y |

## ACTIVITY 8.4

1) T 2) T 3) F 4) F 5) T 6) T 7) T 8) T 9) T 10) T

# CHAPTER 8.5

## ACTIVITY 8.8

1) auditory 2) cervical 3) eye 4) light, deep 5) optic 6) nose 7) cochlea 8) thoracic
9) atlas 10) synesthesia 11) lumbar 12) pain 13) temporal 14) cones 15) temporal

# CHAPTER 9.5

## ACTIVITY 9.6

1) frontal, motor, sensory, parietal, occipital, temporal 2) keeps your heart and lungs going
3) waking /sleeping cycle, wakes you up 4) hypothalamus 5) glial 6) cervical 7) lumbar 8) optic 9) auditory
10) peripheral 11) central 12) involuntary 13) meninges 14) rapid eye movement 15) melatonin 16) beta, alpha
17) delta 18) five 19) 4 20) Samuel Taylor Coleridge

# CHAPTER 10

## ACTIVITY 10.2

1) K 2) E 3) S 4) J 5) M 6) T 7) A 8) X 9) Y 10) U 11) O 12) R 13) N 14) V 15) L 16) F
17) Q 18) W 19) D 20) P 21) H 22) I 23) Z 24) B 25) C 26) G

# CHAPTER 10.5

## ACTIVITY 10.3

Please note that this activity is only meant to be a fun way to review reading material from the chapter. It is not meant to be an actual diagnostic activity!

Patient #1: Alzheimer's
Patient #2: cerebral palsy
Patient #3: epilepsy
Patient #4: stroke
Patient #5: depression
Patient #6: hydrocephalus
Patient #7: tension headaches
Patient #8: concussion

CPSIA information can be obtained
at www.ICGtesting.com
Printed in the USA
BVHW020842150223
658553BV00022B/228

9 798986 863702